Karen S. Kolasa, RD

Dietetic Service Operation Handbook: Practical Applications in Geriatric Care

Pre-publication REVIEWS, COMMENTARIES, EVALUATIONS . . .

"An excellent handbook for dietary service operations. I am especially delighted with the directness with which [the author] addresses each topic. She is very accurate and to the point.

I also appreciate the attention she has given to pureed and dysphagia diets. Swallowing and chewing disorders are very often neglected in dietary references.

THE RECIPES WILL BE A MUCH-NEEDED ADDITION TO OUR RECIPE FILES AND A 'NEW LIFE' TO THE MANY PERSONS ON RESTRICTED DIETS."

Janet Flotemersch, MA, RD, LD
Co-Owner of Dietetic Services of America Consulting Corporation; Clinical Dietitian, Humana Hospital, Lake Cumberland

More pre-publication
REVIEWS, COMMENTARIES, EVALUATIONS . . .

"As a company specializing in the LTC area, we are continually receiving calls for guidance and information concerning nutritional needs, requirements, and documentation. We are confident this book will answer many questions and offer solid solutions to the special nutritional needs of LTC residents.

The book addresses many specialized requirements of the geriatric population. WITH THE GROWING FINDINGS OF THE IMPORTANCE OF NUTRITIONAL ASSESSMENTS AND THEIR IMPACT ON OUR WELL-BEING, IT IS REFRESHING TO FINALLY HAVE A QUALITY, HIGH-CALIBER SOURCE OF INFORMATION FOR REFERENCE. We will highly recommend this book to all of our clients."

Lisa Hanaver
President, Med-Pass Inc.,
Dayton, Ohio

"Broken down into two sections, Administrative and Clinical Topics, this publication PROVIDES THE DETAILED INFORMATION NECESSARY TO MAINTAIN A QUALITY DIETARY DEPARTMENT, AS WELL AS SUPPORTING FACTS AND INFORMATION AS BACKUP MATERIAL. Cross-referenced topics such as Quality Assurance and Medical Records allow the maximum usage out of the material provided by Kolasa. Easy and simple step-by-step procedures in the recipes allow little, if any, room for error on the part of the dietary department but allow the residents the correct amount and necessary nutritional diet."

Christian Haacker, NHA
Nursing Home Administrator,
Bossier City, LA

FPP

Food Products Press
An Imprint of The Haworth Press, Inc.

Dietetic Service Operation Handbook
Practical Applications in Geriatric Care

FOOD PRODUCTS PRESS

New, Recent, and Forthcoming Titles:

Nutrition Care of People with Diabetes Mellitus: A Nutrition Reference for Health Professionals by Penelope S. Easton, Charlotte S. Harker, Catherine E. Higgins, and Marvin C. Mengel

World Food and You by Nan Unklesbay

Yogurt, Yoghurt, Youghourt: An International Cookbook by Linda K. Fuller

Glossary of Vital Terms for the Home Gardener by Robert E. Gough

Dietetic Service Operation Handbook

Practical Applications in Geriatric Care

Karen S. Kolasa, RD

Food Products Press
An Imprint of The Haworth Press, Inc.
New York • London • Norwood (Australia)

Published by

Food Products Press, an imprint of The Haworth Press, Inc., 10 Alice Street, Binghamton, NY 13904-1580

Library of Congress Cataloging-in-Publication Data

Kolasa, Karen S. Thatcher.
Dietetic service operation handbook: practical applications in geriatric care / Karen S. Thatcher Kolasa.
p. cm.
Includes bibliographical references and index.
ISBN 1-56022-027-9 (alk. paper).
1. Nursing homes–Food service–Handbooks, manuals, etc. 2. Aged–Nutrition–Handbooks, manuals, etc. I. Title.
[DNLM: 1. Dietary Services–organization & administration–handbooks. 2. Nutrition–in old age–handbooks. QU 39 K81d]
RA999.F65K65 1993
613.2'084'6–dc20
DNLM/DLC
for Library of Congress 92-49635
CIP

Dedicated to the inspiration
of my grandparents,
Harold C. and Lillie E. Davis,
who taught us that love
is the most perfect gift of all

and

Acknowledging the faithful support and encouragement
of my husband and parents

NOTES FOR PROFESSIONAL LIBRARIANS AND LIBRARY USERS

This is an original book title published by Food Products Press, an imprint of The Haworth Press, Inc. Unless otherwise noted in specific chapters with attribution, materials in this book have not been previously published elsewhere in any format or language.

CONSERVATION AND PRESERVATION NOTES

The paper used in this publication meets the minimum requirements of American National Standard for Information Sciences–Permanence of Paper for Printed Material, ANSI Z39.48-1984.

CONTENTS

CLINICAL TOPICS

ABOUT THE AUTHOR

Karen S. Kolasa, RD, a registered dietitian for 14 years, is the author of several nationally published articles in the field of long-term care dietetics. Her primary interest is in the computerization of seasonal menus, standardized recipes, and practical reports. Her recent work has focused on this area in order to assist food service operations in achieving their maximum potentials so they can provide client satisfaction. Ms. Kolasa is a member of the American Dietetic Association and the Consultant Dietitians in Health Care Facilities Practice Group. Her name appears in the 1993 edition of *Citation's Who's Who Registry of Rising Young Americans*.

Introduction

This manual is a functional tool to initiate effective organization and operation of your dietetic service regardless of its age or condition. It also contains recipes not readily attainable, and forms which are compatible with OBRA laws. When food service employees allocate their time wisely, facilities save time and money, and these employees become more effective members of the interdisciplinary medical team. Any administrator who wishes to control costs through a better comprehension of the dietary department and communication with its employees will gain insight through examination of this text with application to his or her own service.

ADMINISTRATIVE TOPICS

I. Menu Planning

Menus dictate food costs, nutritional adequacy, plate appearance (including contrast of color, texture, flavor, consistency, and shape), type of preparation and equipment, number of employees, food or flavor frequency, garnishes and accompaniments, and client acceptance. However, menus are often not scrutinized for content and accuracy. Instead of being the first priority, they are often considered last! A four-week, nonselect menu cycle usually meets the needs of 50- to 250-bed institutions, but if a select menu is implemented, a two- to three-week menu cycle may suffice. Usually two cycles, a fall/winter and a spring/summer, are sufficient. Generally, five-week menu cycles are not practical due to excess inventory and a low separation rating, i.e., the number of days between times that a menu item is served.[1]

Simplicity, legibility, and clarity are three goals of menu formatting.[2] As the client population changes, the format must grow and adapt with it. It is always preferable to provide extra space in whatever style you use to allow for future developments. Some facilities prefer assigning a number to each menu entree, while others use horizontal and even vertical lines to separate them. The following paradigms may help in deciding which combination is best for you:

SUNDAY WEEK ONE FACILITY'S NAME

Regular	Portion Size	Mech Soft	Puree	1200 Calorie	2 GM. NA
1. Peas	1/2 Cup	1.	1. STR	1. LF	1. LS

SUNDAY WEEK ONE FACILITY'S NAME

Regular	Portion Size	Mech Soft	Puree	1200 Calorie	2 GM. NA
Peas	1/2 Cup	X	STR	LF	LS

Alternative Meat:

Alternative Potato:

Alternative Vegetable:

Nutritional needs of clientele, based on medical condition and physicians' orders, determine the type of therapeutic diets required. Of course, you may want to influence the type and wording of diet orders received by sending physicians a memo or a diet ordering booklet that contains diet names which coincide with the diet manual you have selected. Your diet manual must be functional. Unnecessary special diets increase the propensity for tray errors along with increasing your budget due to extra staff needed to prepare additional foods. The meal pattern calculations for calorie-controlled regimens should be followed. If you prefer to figure your own calculations, they must be included in your policy and procedure manual. (The same low fat milk content is usually planned for regular and calculated caloric diets; nonfat milk is indicated for cholesterol and fat restricted diets.) The meal pattern in Appendix 1 is very functional, using approximately 30 percent total fat content as recommended by the American Heart Association, and approximately 20 percent protein and 50 percent carbohydrate levels as per the American Diabetes Association and The American Dietetic Association. For this pattern, the new "Exchange Lists for Meal Planning" revised in 1989 were implemented.[3]

Try a rotating weekly snack schedule for all diets, emphasizing specialty milk and fruit drink recipes such as the Power Shake in the Specialized Recipes Section. A rich source of Vitamin C is served daily and a Vitamin A-rich vegetable or fruit is served every other day. The menu should meet the Recommended Dietary Allowances for adequacy of every diet, or your diet manual should explain which nutrients do not to meet these requirements. Hint: to meet thiamine recommendations, try serving pork once a week if not contraindicated. Otherwise, this nutrient may be deficient.

The palatability of the foods served is the final measure for menu and recipe writing success. It may be measured by an annual resident survey, daily quality controls such as temperature of food on the tray line and at point of service, attractiveness of plate presentation, and taste. These factors serve as the best evaluation tools. A daily taste panel committee helps to appraise them. It should, of course, include an expert–a client! Why not involve a family or staff member occasionally? The Sensory Evaluation Device provided in Appendix 2 can be used to record their comments about the test tray.

Vary the type of diet tested. A monthly tally of results may be beneficial to your quality control program. A semiannual patient survey conducted by the food service manager helps to determine the general satisfaction of the patient population.

Menus need to be highly visible, especially on the tray line. While there is no substitute for a menu review before meal service, accessibility ensures a greater degree of efficiency. Suspending menus from the ceiling over the tray line, in a plexiglass holder for protection and visibility on each side, is a good suggestion. I prefer them on 11″ × 15″ paper for ease of readability. If there is a quality control checker position at the end of the tray line, he or she definitely needs a legible copy. An explanation of portion sizes and abbreviations, such as those included in Appendix 3, may accompany menus and recipes as a separate posting on the bulletin board. Of course, each menu needs an approval signature and date.

REFERENCE NOTES

1. West, Bessie Brooks et al. *Food Service in Institutions*, Sixth Edition. Philadelphia: MacMillan Publishing, 1988.

2. Kotschevar, Lendal H. and John B. Knight. *Quantity Food Production Planning And Management*, Second Edition. New York: Van Nostrand Reinhold, 1989.

3. American Diabetes Association and The American Dietetic Association. "Exchange Lists for Meal Planning." Revised 1989.

II. Standardized Recipes

A standardized recipe is tested and adapted for use in a given food service operation. It is adjusted according to resident/employee census and contains the name of the food item and file number or code if applicable, total quantity and number of portions of a specific size, ingredients by weight and measure (and sometimes by count), procedures and times for combining ingredients, cooking or baking temperatures and times, panning information, cost information, standard of quality expected, and total time for producing the recipe.[1] Recipe names should be brief and descriptive of the product. Their goal is to assist production of a known quantity of food of a desired quality. Here is a step-by-step process to adjust recipe yields, called the "factor method"[2]:

Step 1. Divide the desired yield by the known yield of the recipe to obtain the basic *factor.*

Step 2. Convert the weight of each ingredient to ounces or parts of ounces as much as possible. In cases where ingredient measures are too small, follow Step 3 carefully. A food substitution guide that measures weight and volume equivalents is a handy reference tool, e.g. the number of cups of chopped carrots in one pound. Your food supplier may also be helpful. (See Appendix 3 also.)

Step 3. Multiply the amount of each ingredient in the original recipe by the factor.

Step 4. Change weights of any ingredients which can be more easily measured by volume than by weight back to volume. Convert to largest units.[1]

Step 5. Document the end results in pencil on your recipes because your census and production needs vary.

REFERENCE NOTES

1. Kotschevar, Lendal H. and John B. Knight. *Quantity Food Production Planning and Management*, Second Edition. New York: Van Nostrand Reinhold, 1989.

2. The American Dietetic Association. *Standardizing Recipes for Institutional Use*. Chicago: The American Dietetic Association, 1967.

III. Food Purchasing and Production

Purchasing is a logical process. Always request a proposed pricing from all suppliers within a specific time frame, e.g., a 30-day cycle whereby prices go into effect on the first of the month and remain in place until the end. If possible, ask suppliers for a list of all mark-ups by cost per unit or percentage. When negotiating a bid, ask suppliers to explain the formulas used to determine their prices. Inspect items by weight or count and date code upon arrival in determining their quality. Always write the current date on each item before storage. Here is a suggested logical method for routine ordering:

Step 1. Review the entire week's menu, listing all items that are required.

Step 2. Design a sheet showing the quantity of an item used for a meal. For example, at lunch you may use two #10 cans of regular fruit. By using your ordering guides, and the "Contents of Cans" list in Appendix 3 (which shows that one #10 can contains 12 cups), you can calculate the amount of the item required to meet your needs.

Step 3. Before you begin ordering, take the list of items made in Step 1 into your storage areas and count how much of each item you have; record these counts on your list or inventory book. (Record all counts by the number of smallest units, such as cans, not case lots.)

Step 4. Next, check incoming items and add these amounts to the numbers on your list from Step 1.

Step 5. Using your menu extensions which show the entire weeks' menus for the current week, the week following, and the week for which you are ordering (assumed to be two weeks in advance), begin subtracting the number of items you need for each day from the total number you have in stock.

The following is an example; it assumes that there are five cans in stock, that you use two cans per meal, and that you order two weeks in advance:

> Regular peaches appear one more time on the menu you are currently serving, once on the next week's menu, and twice on the menu of the week for which you are ordering. Therefore, with five cans in stock, less two cans for the week you are serving and two cans for next week, one can would be left. You will need a total of four cans for the week for which you are ordering (two #10 cans times two days served). Thus, three more cans are required. If you have established a par or minimum level to keep on hand, this order may vary.

Follow this method for each item on your list. It takes about one to two hours each week to accomplish a thorough, accurate grocery order.

Careful record-keeping in your storeroom also makes detection of pilferage possible. Keep a weekly record of items in your pantry and refrigerated areas. This ensures stock will be used on a rotational basis (first-in, first-out), prevents food spoilage, and thus controls unnecessary spending. The bottom line remains client and customer satisfaction. These factors must be reviewed with continual product and service reevaluations, especially to achieve minimal plate waste and maximum sales.

Many facilities feed both patients and staff. For these busy kitchens, a production sheet becomes essential. The production sheet has a two-fold purpose: it serves as a daily work guide for employees and assists the managers in their menu predictions.

The production sheet or schedule serves as the work sheet for the day and is the best way for the manager to communicate to every employee exactly what he/she expects for this particular day. Each employee is given the same information at the same time. It also relieves the employee of the responsibility of making too much or not enough of a certain menu item. If a cook prepares 150 servings of a recipe and only 50 servings are used, then the surplus is the manager's responsibility, not the cook's. However, if a cook grills 25 more sandwiches than stated on the production sheet and has 30 sandwiches left over, it is the cook's responsibility that the food and labor have been wasted.

The production sheet also can be an aid to forecasting. One of the hardest tasks a manager has is forecasting how much of each item clients and staff are going to want for any given day. If the production sheet has been properly filled in, it will tell the manager how much of each item was used the last time this particular combination was served. The manager can then forecast how much of each menu item to order and prepare.

From past production sheets, the manager knows how many clients will stay in the facility for a particular weekend or holiday and can adjust production accordingly. The menu can be changed for control items at these times to avoid overproduction.

If the production sheet has been properly filled in the previous day, the manager can easily tell what the leftovers are and where they are stored. If they are in the refrigerator, they should be removed and used at the following meal if possible. If a leftover has been frozen, the manager can decide when to use the item and reduce future requisition orders.

The production sheet also assists the manager by listing the following information: location of recipe to save time in finding it; portion size for standard servings; line setup time to tell the employee at what time the line should be clean and ready to use; and the following day's production, listing items that need to be prepared the day before served or items that can be prepared ahead.[1,2]

REFERENCE NOTES

1. West, Bessie Brooks et al. *Food Service in Institutions*, Sixth Edition. Philadelphia: Macmillan Publishing, 1988.

2. Kotschevar, Lendal H. and Charles Levinson. *Quantity Food Purchasing*, Third Edition. New York: Van Nostrand Reinhold, 1988.

IV. Equipment Care

Modern and suitable equipment in good working condition is essential for control in management. The elimination of some of the human element leads to greater efficiency and to more standardization of output and quality. This needs in no way to detract from the attractiveness of the product. By eliminating many inefficiencies and "margins of error," more time can be devoted to achieving high food quality and the flair of garnish and service.

The efficiency of any machine will depend on its wise selection, use, care, and upkeep, especially daily maintenance. The need for the machine must be justified and knowledge of its performance should be acquired before any selection is made.[1] In fact, any equipment that will cut labor hours and production time must be considered. For example: a commercial potato peeler with centrifugal force operation has a specialized function and costs around $1000, which may not fit into your budget; however, an automatic fruit and vegetable peeler such as the DAZEY Stripper® would be more versatile and costs under $50. Preventative maintenance is essential, and the wise manager will schedule a complete in-service with the manufacturing representative to cover the operation and ongoing care of equipment. Always inspect the equipment before the warranty expires in order to gain maximum service before additional costs apply. All written manuals should be kept handy. See Appendix 4 for a proposed "Sanitation Checklist." Initial and ongoing attention to proper equipment care is an investment in its sound operation and pays dividends through reduced repair costs, especially after the warranty expires.

REFERENCE NOTE

1. Kotschevar, Lendal H. and Margaret E. Terrell. *Food Service Planning: Layout and Equipment*, Third Edition. Philadelphia: Macmillan Publishing, 1985.

V. The Policy and Procedure Manual

A policy and procedure manual should reflect the current operation of the Dietary Department, as well as its coordination with other departments including Nursing, Maintenance, Activities, and Housekeeping. The manual should be updated at least annually by supervisory staff and consultants directly involved with the nutritional care of residents. All dietary employees should be familiar with its contents. It should never be simply a restatement of the regulations, nor be limited to only those broad areas identified in regulations, nor be a compilation of materials from reference manuals. It should address at least the following areas:

1. Statement of purpose of department.
2. Budget and purchasing: accountability, frequency, record-keeping, responsibilities.
3. Staffing: organizational chart; supervision (hiring, orientation, scheduling, evaluations); job descriptions and work schedules; in-service (frequency, annual plan, records, programs for non-dietary staff).
4. Menus: type and length of cycle, e.g., a selective, four-week; meal pattern and adequacy; development, approval, revision, and seasonal changes; posting and retention; supplements (see "Example: Nutritional Supplementation Guidelines" that follows); holiday menus.
5. Therapeutic diets: types and objectives; clarification, transmittal, and update of orders; responsibility for consistency changes; menu development and approval; diet manual and adaptations such as house diets; records kept in dietary; supervision of service; holiday practices.
6. Food preparation: recipes (use, yield adjustments, nutrient analysis); tasting; methods for blending, grinding; use, handling, and preparation of raw, frozen, and convenience items;

meat and vegetable cookery; production sheets; portion control.

7. Meal and nourishment services: tray and/or dining room service; times and responsibilities for assembly, delivery, supervision/assistance/monitoring intake; tray and snack identification; adaptive equipment; replacements and substitutions; use and ordering of commercial supplements/tube feedings; supplies for other departments such as Activities; employee meals or guest trays.
8. Sanitation: food preparation, handling, and storage; hygiene of staff; waste disposal; pest control; testing and holding temperatures; use and testing of sanitizing agent; labeling and dating items; leftovers.
9. Equipment and physical environment: correct operation; cleaning and maintenance schedules (item, frequency, method [including amounts of soap/sanitizer], personnel); reporting malfunctions; ventilation.
10. Infection control: employee health (annual and incidental); isolation methods; hand washing; food handling; use of sanitizer temperatures.
11. Charting: time frames and responsibilities; forms used; methods for visitation, implementation, coordination, evaluation, and referral.
12. Miscellaneous: emergency feeding plan (menu, supplies, potable water, loss of utilities); safety; security; residents' rights (preferences, complaint resolution, refusal of treatment); uniform or dress code.

Local and state health departments can provide excellent advice in these areas and others regarding their rules and regulations. Subjects to cover with them include: the topics to be discussed at dietary in-service meetings, and how frequently; recommended number of therapeutic diets and sugestions for specific types; and any written materials they distribute to assist in employee training or staff supervision. In-service planning may be facilitated using the forms in Appendixes 5 and 6 (both of which may be used for advanced posting on an annual or weekly basis). Suggested topics include any of the above mentioned areas plus personal hygiene, cardiac diets, diabetes, sodium restricted diets, proper food storage and labeling, and correct use of plastic gloves.

Monthly consultant report forms, such as Appendix 7, summarize documentation performance for the consulting dietitian and indicate short- and long-term goals to key members of the dietary staff, the Director of Nursing, and the Administrator. The consulting dietitian usually makes it a point to stop at the Administrator's office before beginning or leaving work, if possible. The dietitian completes a minimum of one annual in-service for the entire staff. The nature of this position is outlined in the Position Description that follows. You may wish to borrow from this description when drawing up the dietitian's mandatory contract. A combined job description/performance evaluation is a time-saving tool. After each duty, provide space for comment.

Example: Nutritional Supplementation Guidelines

Order Name: By type and amount of supplement.

Indications for Usage: Supplements provide extra nutrients sufficient to meet the metabolic needs of individuals whose requirements are greater than those provided by standard meal and snack plans. This could be due to inability to consume large enough amounts of food at one time.

Characteristics: Nutritional supplements can be either foods that are added by the diet technician/dietitian, or supplements that are prescribed for an individual's diet by his or her physician.

Comments: Commercial products are an option for those individuals not receptive to efforts to increase oral nutrient intakes through foodstuffs, or who cannot consume the prescribed amount of food. However, foods are the most desired form of nutritional supplementation because they encourage normal eating patterns and are more cost effective than commercial products. (The advanced kitchen may wish to prepare snacks using recipes provided by pharmaceutical companies and which contain their products.) Examples include but are not limited to: milk shakes, cheese, crackers, custards or puddings, sandwiches, cottage cheese, peanut butter, cereal, milk, juice.

Position Description

Title: Consulting Dietitian
Approved By: ________________
Principal Duties and Responsibilities:

1. Reviews short- and long-range department plans and programs consistent with in-house policies, and revises and updates policy and procedure manual.
2. Establishes and maintains standards for food preparation and service, sanitation, housekeeping, safety, and security.
3. Observes facility routinely and evaluates to ensure that all objectives and responsibilities are met.
4. Recommends suppliers and participates in the bid process.
5. Develops specifications for food, supplies, and equipment to be purchased.
6. Assists in the planning and evaluation of orientation and in-service educational programs, and conducts an annual interdepartmental program for all staff.
7. Interprets, evaluates, and utilizes pertinent current research relating to nutritional care and dietary management.
8. Confers with the Administrator, Director of Nursing, Physicians, and Dietary Manager on a regular basis.
9. Maintains effective written and verbal communications and public relations.
10. Develops and implements forms and patient education materials to facilitate maximum continual clinical and administrative quality improvement.
11. Instructs clients on complicated diet regimens.
12. Approves menu patterns and assists in evaluating clientele acceptance.
13. Routinely documents all high-risk patients' medical records.
14. Submits a monthly written report to the Administrator.
15. Advises the food service manager based on pertinent findings.

VI. Quality Improvement

Every department needs a quality management program. If there is a quality assurance coordinator, he or she is the resource person in terms of setting up specific programs for the entire facility. A ten-step approach is widely recognized as the acceptable method to organize and monitor your professional standards. Actual outcome in terms of better resident satisfaction is the ultimate goal. The ten steps are:

1. Assignment of responsibility
2. Delineation of the scope of care or service
3. Identification of important aspects of care or service
4. Identificaiton of indicators or measurable variables
5. Establishment of thresholds for evaluation
6. Collection and organization of data
7. Evaluation of the data
8. Action to resolve identified problems
9. Assessment of actions and documentation of improvement
10. Communication of relevant information to the QA Coordinator.[1]

Possible dietetic indicators are appropriateness of diet orders, including time span of NPO (No Pass Oral) or clear liquid diets, and timeliness of nutritional assessments and progress notes (see Appendix 8). The following presents these aspects of service through the various stages:

QA Responsibility: ________________ (name of person responsible from dietary)

Scope of Care: Dietetic Services will provide nutritionally adequate and balanced meals to all of the patients at ________________ (facility name).

Aspects of Care: Dietary will document the appropriateness of the diet orders as received by the wards. Examples: (1) Mechanical soft diet will be ordered instead of ground meat diet; (2) Modified fat/low cholesterol diet will be ordered instead of low cholesterol diet.

Indicators: Clients on NPO or on unsupplemented clear or full liquid diets for more than 48 hours should be receiving some type of nutritional support. Patients with diagnosed medical problems such as diabetes, hypertension, heart problems, or other special medical needs will be recorded on the Nutrition Screening form (see Appendix 9) and will be evaluated within five working days. Patients who have a cholesterol reading above 239 mg/dl will be recorded on the Nutrition Screening form and will be evaluated within five working days. All patients will have an annual nutritional assessment, and will have a follow-up evaluation on a quarterly basis.

Collection of Data: The Nutrition Screening form will be used at admission on all wards to prioritize the need for nutrition intervention. Previous assessments, progress notes, and care plans from the patient medical record will be used.

Thresholds for Evaluation: One-hundred percent of clients on NPO or on unsupplemented clear or full liquid diets will receive some type of nutritional support. Ninety-five percent of patients who have other medical problems will have initial and annual nutrition assessment completed, and quarterly progress notes and plan of care will be updated. Ninety percent of patients who have a cholesterol reading above 239 mg/dl will have an annual nutrition assessment completed, and quarterly progress notes and plan of care will be updated. Ninety percent of all clients will have an annual nutrition assessment completed, and quarterly progress notes and plan of care will be updated.

Evaluation: The Nutrition Screening form to identify clients at-risk will be prepared by the dietitian. These forms will be kept on all wards. The dietitian will prepare initial nutrition assessments and follow-up on those clients who require nutritional intervention.

Action: The Nutrition Screening form is a major tool used to evaluate the components of the QA Program. The dietitian will update the QA Program annually.

Assessment of Actions and Documentation of Improvement: If initial evaluation and action to correct the QA problem does not produce results, then a secondary evaluation will be made.

Communication of Relevant Information to the QA Coordinator: A Monthly Summary Work Sheet (example follows) will be sent to the QA Coordinator. Remember, each indicator must have corresponding threshold(s) for evaluation.

MONTHLY SUMMARY WORK SHEET

INDICATOR	MET	NOT MET	%	COMMENTS
Unsupplemented diet < 48 hours				
Medical problem met w/in 5 days				
Chol > 239 mg/dl met w/in 5 days				
Nutrition notes Assessments– Initial/Annual				
Quarterly– Progress notes				
Plan of Care				
% is the number met divided by the total number done for each indicator. Other indicators to consider for monitoring are transcription of diet orders, identification and accommodation of food allergies, accurateness of special diets, food portion adequacy, plate waste, and availability of diet counseling.				

ACKNOWLEDGEMENT

The Nutrition Screening Initiative (NSI; launched in May 1990) is a five-year, multifaceted effort to promote routine nutrition screening and better nutrition care in our health care system. The initial focus is on the elderly, one of the largest groups of Americans at risk for poor nutrition.

REFERENCE NOTE

1. The Joint Commission on Accreditation of Healthcare Organizations. "Monitoring and Evaluation: Physical Rehabilitation Services." Chicago: The Joint Commission on Accreditation of Healthcare Organizations, 1987.

VII. The Pureed Process

The French-derived term "puree" literally means to strain (not liquefy) or to make pure. Ironically, when applied to food production and service, this diet tends to be the most difficult and time-consuming of all to prepare and serve. Devitalized leftovers, mixed in unclean blenders or processors, held at a dangerous temperature, and served as runny textures, do not present an appetizing or nutritious picture, nor do they appeal to any palate. The goal is to provide a consistent texture from one day to the next.

EQUIPMENT REQUIREMENTS

The food processor, blender, and cutting board are the pieces of equipment most frequently associated with the puree process. The facility will save both time and money by purchasing its own commercial food processor. Follow the operations manual carefully for maximum benefit in the areas of maintenance, sanitation, and safety for all equipment. Here are a few general rules:

1. Most manufacturers do not recommend cleaning the processing containers or covers in the dishwasher. Therefore, wash those pieces with detergent and sanitize for at least two minutes in a rinse solution. The base needs to be wiped down with the rinse solution also. (Rinse solution refers to the proper amount of water added to a sanitizer as indicated on its label.) Since the blades and removable parts must be cleaned after each use to meet federal standards, this process must become routine.
2. Similar rules apply to the blender; however, this container should not be immersed in water. Fill the container with hot tap water to cover the blades, add detergent and cover. Press two consecutive buttons of increasing speeds to reach full speed. Rinse, allowing for the two-minute contact time with the sanitizer. Never place boiling hot liquids in a glass blender!

3. Always unplug electrical equipment before cleaning.
4. Never immerse an electrical motor in water.
5. Use separate cutting boards for raw and cooked foods. Scrub them in hot, soapy water immediately after each use. Rinse under hot running water. (Many facilities prefer the easier-to-clean plastic or rubber boards even though the hard/wood boards last longer.)

Some food processor attachments may be safely cleaned in the top shelf of the dishwasher. Refer to your manual for further usage rules.

Sanitized equipment is essential before blenderizing. Begin with ingredients prepared and/or stored using good sanitation techniques. No foods should ever reach the temperature danger zone of 40° F to 140° F, to prevent food-borne illnesses. After blending, place food in a steam table pan of proper size. Heat the food in a steamer using plastic wrap to cover, or heat in a conventional oven using aluminum foil to cover. Regardless of the heating method used, the mixture must reach 160-180° F. Precooked foods should be held in a preheated steam table (or if a cold food, held in a chilled compartment) no more than 20 minutes prior to serving. Promptly refrigerate or freeze any leftovers, wrapping them securely, to prevent spoilage. Generally, refrigerated items should be used within 24-36 hours. (For unopened, preportioned food items such as cartons of milk or juice, etc., defer to their expiration date.) Thaw frozen entrees using one of two methods: either by placing them in the refrigerator or under cold running tap water prior to use.

With the advent of the food processor, cooks are spared the tedious task of constant scraping with liquid additions in preparing entrees of a smooth consistency. An institutional food processor may prove more durable than a blender in the long run and require less liquid additions thereby reducing the amount of thickener needed. Chefs usually prefer "pulsing" of the blades to promote even distribution of the product.

PREPARATION METHODS FOR PUREED FOODS

Puree implies that the food should be of a mashed potato-like consistency. Just as potatoes should be peeled, cooked, and completely mashed for ultimate creaminess before the addition of liquids such as milk, the same is true for other pureed entrees. Consider the following categories:

Meats and Casseroles

1. Puree hot entree at a low speed into a paste, to remove any grainy texture before adding the liquid. Usually 1/4 cup liquid, such as broth, is added to 1 cup diced meat.
2. Gradually add warm liquid, pureeing to reach a soft consistency, e.g., try 2 Tbsp. liquid per 3/4 cup casserole or 3 oz. meat portion.
3. Increase the cohesiveness by adding small amounts of a stabilizer–instant potato, non-fat dry milk powder, bread crumbs, toast, or commercial thickeners. For example, add one slice of bread, or 1-2 Tbsp. bread crumbs, instant dry potatoes, hot cereals, or non-fat dry milk per 3 oz. or 1/2-cup portion. (A stabilizer is any agent that adds body, e.g., unflavored gelatin, which absorbs water and gives a smooth texture.)
4. Puree on high speed until it hold its shape much like meringue.
5. Some meats such as sausage require both instant potato and non-fat dry milk as the stabilizers. Casseroles may not require a stabilizer. The denser the meat, the less thickener needed or vice versa.

Vegetables, Salads, and Soups

1. Drain completely. Again, about 1/4 cup of drained liquids, etc., to 1 cup solids is an adequate proportion.
2. Puree on high speed, and add liquid and stabilizer when necessary. (Crackers are a good stabilizer for soups as are bread crumbs for tossed salad.) Follow the thickening ratio listed above. Dried legumes (especially navy and split pea) have great application in thick soup purees. The $\propto$-galactosides responsible for flatulence may be reduced by soaking and cooking then discarding the water. Note: a blender is preferred for soups.

Breads

Quick breads or muffins may need only soaking. Usually 2-4 Tbsp. liquid per serving yields the desired texture.

Desserts and Fruits

1. Add approximately 2-4 Tbsp. liquid per serving to pies, brownies, or cookies for the appropriate consistency. Plain, iced cakes may not need soaking.
2. Only 1 tsp. liquid for 1/2 to 3/4 cup fruit is necessary, regardless of whether it is fresh, canned, or frozen.

PRESENTATION AND SERVING SUGGESTIONS

Proper presentation includes serviceware that is sized for the items displayed. Side dishes or even sectioned plates are preferred to hold the food's shape. Main plates (not bowls) are fine for casserole dishes. Do not be afraid to bring out the stemmed glassware to present desserts or beverages. Try to practice truth in advertising. If the natural color of a food item is green, keep it green! Instead of blending gravies or sauces into entrees, ladle them over the items for decoration.

A garnish is an edible accent added to a dish to decorate one food with another. It may also enhance caloric density. Seasonings and flavorings, although noncaloric, add greatly to color and flavor. Other accents give texture variations. The Specialized Recipes section that follows includes helpful garnishing suggestions for variety and contrast. A garnish may also be used within the food itself. To promote a greater number of calories in each bite, try the following ideas:

1. Use whole milk (instead of skim or 2%), half and half, or undiluted evaporated milk to prepare soups, puddings, or hot cocoa, or use Fortified Milk (recipe p. 71) in cooking.
2. Try sour cream in mashed potatoes, blended meats, and fruits.
3. Adding 1/4 cup dry powdered coffee creamer to every 1 cup serving of milk shakes, hot cocoa, hot cereal, gravies, and soups increases the calories by 170 for each portion.
4. Whipped cream on desserts or folded into puddings tastes great, too! Serve fruit packed in heavy syrup as a salad or dessert.

5. Eggnog instead of milk with farina is delicious. Cook hot cereals with milk instead of water.
6. Smooth peanut butter mixed with additional butter may be tolerated by some patients; also try adding 1 Tbsp. of peanut butter to 1/2 cup of cream of celery soup.
7. Offer extra sugar and margarine whenever possible.
8. Prepare gelatin using fruit juice instead of water.

As in all menu planning, the aesthetic values of color, texture, aroma, and flavor apply. For instance, do not use more than one gelatin dish even if it is another entree category, and never repeat a color too frequently. Seasonal cycle menus (fall/winter and spring/summer) determine the types of pureed foods you will serve. The rule of thumb is to plan menus to utilize as many "across the board" items as possible. Pureed-like foods such as aspics, plain ice creams, sherbets, sorbet, frappes, ices, frozen yogurt, gravies, sauces, glazes, and toppings, and beverages such as milk, smooth milk shakes, and fruit juices appeal to everyone.

Here is a suggested menu format for purees:

Breakfast	*Dinner*	*Supper*
Fruit Juice or Puree	Fruit Juice	Same as Dinner except add Soup and delete Fruit Juice.
Egg Dish	Meat or Casserole Dish	
Hot Cereal	Potato or Substitute (Gravy)	
Toast or Bread		
Margarine and Jam or Jelly	Vegetable (Sauce) Fruit (Glaze)	
Milk	Dessert	
	Beverage	

Obviously, you will vary the usage of gravies, sauces, etc., just as soup will alternate with fruit juice. Some patients may not require pureed meats; therefore, a combination diet order such as "pureed with ground meat" must be secured. Various combinations are possible, e.g., "pureed with regular fruits and vegetables." The goal is to serve 2000-2500 calories daily to most of these clients.

FURTHER DIETARY CONSIDERATIONS

Typically, the potential for weight and protein loss are major risk factors with the clientele on this regimen. Patients who ingest fewer calories and protein because of the high liquid content of their diets are in serious danger. When patients have pressure ulcers or other conditions that require extra nutrients, the need to intervene increases. Any situation that reduces protein intake, thus potentiating a negative nitrogen balance, lessens the potential for drug therapy efficacy. Serving broths, stocks, juices, milks, and creams (instead of just water) with meats, vegetables, and fruits, and even serving larger portions, facilitates higher caloric intakes. When purees are prepared and served properly, the use of commercial supplements should be minimized. When a commercial supplement seems mandatory, remember to select a nutrient-dense formula, i.e., obtain maximum energy nutrients per volume, such as pudding. Commercial products formulated for renal patients are useful when fluid, potassium, and sodium restrictions apply. Physicians' orders must be followed carefully when calculating supplements for diabetics to keep within dietary guidelines. Pharmaceutical companies supply product handbooks which are excellent resources for special supplementation.

In the Specialized Recipes section that follows, note that the caloric distribution and dietary fiber contents are significant. The addition of oat bran and Miller's bran plays an especially important role in promoting bowel motility, which may be reduced by the texture alteration of the diet and by the likely low physical mobility status of recipients consuming the pureed diet.

Be creative, but more importantly, experiment before the actual event to guarantee success! The proof is still in your pudding.

VIII. Specialized Recipes

It is rare when a section of any food service handbook is totally devoted to pureed foods. Presented here are basic entrees with emphasis on Main Dishes, Vegetables, Fruits, Salads, Desserts, Sauces and Supplements. It is instructional to mention that the Standardized Recipes information (discussed in Chapter II) is part of culinary science, and the following examples are no exception. For your convenience, the "Adapted to ____ Servings" column allows adjustment to your census needs. (You may wish to pencil in your figures underneath this heading in the rows corresponding to the ingredients.) I encourage you to determine cost/serving, quality expectations, and preparation time as dictated by your managerial expertise. If the item is appropriate for calorie-controlled diets, exchanges are calculated. A Glossary of Food Terms is included at the end of this chapter. Please refer to Appendix 3 for explanation of abbreviations and equivalents.

Bon Appétit!

MAIN DISHES

Recipe Name: Pureed Cooked Meat (beef, chicken, ham, turkey, etc.)
Portion Size: 1/4 cup
Servings: 10

Adapted to ___ Servings

1 1/2	lbs.	Cooked Meat
1-1 1/2	cups	Hot Stock
1/2-3/4	cup	Whipped Potatoes

1. Grind meat before blending. Meat must still be hot or warm before blending.
2. Blend meat 3-4 minutes, gradually adding enough hot stock to achieve uniform texture and smooth consistency.
3. Scrape meat down. Check with scoop for thickness. Add enough whipped potatoes to hold meat together to scoop.
4. Serve with #16 scoop over mashed sweet potato, bread, pureed noodles, etc. Use 1 oz. ladle of gravy.

Nutrient Content/Serving:			*With Gravy:*	
Calories:	182		227	
Protein:	17	grams	17	grams
Carbohydrate:	1.5	grams	1.5	grams
Fat:	12	grams	17	grams
Fiber:	0.46	gram	0.46	gram

Hint: A sauce with sugar may not be appropriate for some diabetics and would probably not equal one fat exchange. (For recipes, see Sauces section.)

Equals two meat exchanges. Add one fat exchange with gravy.

Garnish: Pureed fruits, chilled or heated.

Recipe Name: Roast Beef or Pork
Portion Size: 2 oz.
Servings: 16

Adapted to ___ Servings

2	lbs.	Top or Bottom Round Beef Roast, or Loin-end or Boston Butt Pork Roast
1	Tbsp.	All Seasoning Salt
2	tsp.	Caraway Seeds or Marjoram
1/8	tsp.	Garlic Powder (optional)

1. Preheat conventional oven to 350° F. Place roast in small baking pan.
2. Rub All Seasoning Salt, caraway seeds or marjoram, and garlic powder (optional) on all sides of roast.
3. For best flavor, cover and refrigerate overnight.
4. Cook uncovered (dry heat) approximately 40 minutes per pound.
5. When meat is done, allow to partially cool; remove from pan.
6. Pour 1/2 cup hot water in pan to dissolve the drippings, then add enough water to drippings to make 2 cups. Use this to make brown gravy.
7. Cut meat into chunks and place in food processor. Process until very fine consistency. Add natural juices or brown gravy 1/4 cup at a time until consistency of whipped potatoes.
8. Serve with #16 disher. Pour brown gravy over meat.

Nutrient Content/Serving:			*With Gravy:*	
Calories:	146		191	
Protein:	14	grams	14	grams
Carbohydrate:	0		0	
Fat:	10	grams	15	grams
Fiber:	0		0	

Equals two meat exchanges. Add one fat exchange with 1 oz. gravy.

Garnish: Applesauce in souffle cup.

Hint: Meat may be served over whipped potatoes.

Recipe Name: Baked Chicken or Turkey
Portion Size: 2 oz.
Servings: 24

Adapted to ___ Servings

3	lbs.	Chicken (whole/fryer)
2	tsp.	All Seasoning Salt
1	tsp.	Summer Savory
1	tsp.	Tumeric

1. Preheat conventional oven to 350° F (convection oven to 275° F). Place chicken in square pan 9″ × 9″ × 2″.
2. Sprinkle seasonings over entire chicken surface.
3. Place chicken, uncovered, in oven and bake for approximately 1 hour.
4. After baking, remove skin and bones.
5. Place meat in food processor and process until very fine.
6. Add chicken broth or gravy 1/4 cup at a time until consistency of mashed potatoes.
7. Serve with #16 disher. Pour 1 oz. gravy over poultry (or serve with barbecue sauce or sweet and sour sauce).

Nutrient Content/Serving:			*With Gravy:*	
Calories:	131		176	
Protein:	15.5	grams	15.5	grams
Carbohydrate:	0		0	
Fat	7.7	grams	12.7	grams
Fiber:	0		0	

Equals two meat exchanges. Add one fat exchange with gravy.

Garnish: Gravy or sauces.

Hint: Some like poultry served over a pureed chicken stuffing.

Recipe Name: Baked Fish
Portion Size: 2 oz.
Servings: 5

Adapted to ___ Servings

1	lb.	Frozen Fish Fillet
1/4	tsp.	Thyme
1/2	tsp.	Dillweed
1/4	tsp.	Marjoram
1	tsp.	All Seasoning Salt
1	cup	2% Milk

1. Preheat conventional oven to 350° F (convection oven to 275° F). Place frozen fish fillet in 2-inch deep baking pan.
2. Sprinkle seasonings on top of fish. Pour milk over.
3. Bake for 45 minutes in conventional oven (30 minutes in convection oven).
4. Puree. Serve with #16 scoop.

Nutrient Content/Serving:

Calories:	196	
Protein:	24	grams
Carbohydrate:	2.4	grams
Fat:	10	grams
Fiber:	negligible	

Equals three lean meat exchanges. If less than 1 Tbsp. of melted butter is drizzled on 2 oz. serving, equals three medium-fat meat exhanges.

Garnish: Spices as indicated.

Recipe Name: Cheesy Vegetable Quiche (crustless)
Portion Size: 1 slice
Servings: 6

Adapted to ___ Servings

1	cup	Swiss Cheese
1	cup	Cheddar Cheese
1/4	cup	Parmesan Cheese
2	cups	Spicy Carrot Puree (see recipe p. 45)
6	ea.	Eggs
1	cup	2% Milk
1	cup	Mashed Potatoes
1/4	tsp.	Nutmeg (optional)
1/2	tsp.	Salt
1/4	tsp.	White Pepper
1	Tbsp.	Cornstarch

1. Preheat conventional oven to 350° F (convection oven to 275° F). Prepare 10-inch pie pan. Spray with vegetable oil.
2. Measure first 5 ingredients into food processor. Process until smooth consistency. Whisk together.
3. Add remaining ingredients and process until smooth.
4. Pour into greased 10-inch pie pan.
5. Bake 50 minutes or until knife inserted in center comes out clean.
6. Allow to set for 5 to 10 minutes.
7. Cut into 6 wedges. Serve hot.

Nutrient Content/Serving:

Calories:	227	
Protein:	18.7	grams
Carbohydrate:	8.1	grams
Fat:	13.3	grams
Fiber:	2	grams

This nutritious main dish equals two meat exchanges, 1/4 2% milk exchange, and and one vegetable exchange.

Garnish: Jelly.

Recipe Name: Cheese Strata
Portion Size: 1 slice
Servings: 15

Adapted to ___ Servings

15	sl.	Wheat Bread
1/2	lb.	Shredded Cheese (any kind)
24	ea.	Eggs (large)
3	cups	Whole Milk
2-3	Tbsp.	Herb or Spice, such as basil, oregano, parsley, pepper
as needed		Paprika

1. Preheat conventional oven to 350° F. Coat a 12 × 20 × 2-inch steam table pan with vegetable spray. Line prepared steam table pan with bread slices.
2. Sprinkle shredded cheese over bread.
3. Beat together eggs, milk, and seasonings.
4. Pour egg mixture over bread; sprinkle with paprika. Cover and refrigerate several hours or overnight.
5. Bake, covered, for 45 minutes. Uncover; continue baking 20-25 minutes, until puffy and golden brown. Let stand 5 minutes before serving. Cut into 15 servings.

Nutrient Content/Serving:

Calories:	431	
Protein:	27.6	grams
Carbohydrate:	16.1	grams
Fat:	28.5	grams
Fiber:	1.3	grams

Garnish: Ketchup in souffle cup.

Note: Great for dysphagics who cannot tolerate bread, but substitute white bread for wheat bread.

Recipe Name: Pureed Spaghetti and Meatballs with Sauce
Portion Size: 3/4 cup
Servings: 10

Adapted to ___ Servings

4	cups	Hot Cooked Spaghetti (1 1/2 lbs. cooked weight)
1	cup	Hot Skim Milk
1 1/4	lb.	Cooked Ground Beef
1 1/4	cup	Skim Milk or Stock
3/4	cup	Whipped Potatoes
2	cups	Spaghetti Sauce

1. Process hot spaghetti with skim milk or stock, gradually adding enough milk/stock for proper consistency, approximately 3-4 minutes.
2. Blend meat for 3-4 minutes, gradually adding enough milk or stock to achieve uniform texture and smooth consistency.
3. Check with scoop for thickness. Add enough whipped potatoes to hold meat together with scoop.
4. Using 1 1/2-2 Tbsp. of meat for each meatball, roll into shape. Serve two each with a # 16 scoop of the pureed spaghetti. Top with 3 oz. spaghetti sauce.

Nutrient Content/Serving:

	Spaghetti:		*With Meatballs and Sauce:*	
Calories:	37		246	
Protein:	2	grams	19	grams
Carbohydrate:	7.2	grams	20	grams
Fat:		trace	10	grams
Fiber:	0.64	gram	1.1	grams

Equals two meat exchanges, one starch exchange, and one vegetable exchange with meatballs and sauce.

Garnish: Italian blended seasonings.

Note: Any type of pasta can be substituted for spaghetti noodles.

Recipe Name: Orange Roughy with Lemon Sauce
Portion Size: 3 oz.
Servings: 24

Adapted to ___ Servings

For Sauce:	3/4	cup	Water
	1/2	cup	Margarine
	1/4	cup	Lemon Juice
	1	Tbsp.	Cornstarch
	3	lbs.	Orange Roughy Fillets
	1	quart	Water
	3	Tbsp.	Lemon Juice
	2 1/4	cups	Egg Whites

1. To prepare sauce: Mix together water, margarine, lemon juice, and cornstarch. Cook over medium heat until mixture is clear, stirring constantly.
2. Bake fillets at 350° until no longer translucent (about 10 minutes per inch of thickness). Quickly chill to below 40° F. Chop fine. Place water and lemon juice in bowl of food processor. Gradually add fish, processing until smooth. Add egg whites; mix well. Pour mixture into well greased, 2 1/2-inch deep, half-size steam table pan. Refrigerate until ready to cook.
3. Steam or bake, covered, at 350° F until mixture is set, about 50 minutes. Let set for 5 minutes before cutting. Serve hot with 1 oz. of sauce.

Nutrient Content/Serving:

Calories:	117	
Protein:	10.7	grams
Carbohydrate:	1	gram
Fat:	7	grams
Fiber:	0	

Equals almost two meat exchanges without the sauce. A generous dollop of cottage cheese brings this delicious entree to an even two meat exchanges.

Garnish: Beet relish.

Hint: For cheese sauce, refer to recipe in Sauces section.

Recipe Name: Ham Salad Loaf
Portion Size: 1/10 loaf
Servings: 10

Adapted to ___ Servings

1	lb.	Precooked Ham, cut into chunks
2	ribs	Celery
2	ea.	Hard Cooked Eggs
1/2	cup	Mayonnaise
3	env.	Unflavored Gelatin (1 env. = 1 Tbsp.)
1/4	cup	Cold Water
1/2	cup	Boiling Water

1. Spray loaf pan with vegetable oil.
2. Pour cold water over gelatin and set aside to soak.
3. Place ham, celery, and hard cooked eggs in food processor and blend until smooth consistency.
4. Add mayonnaise to ham mixture and continue processing until well blended.
5. Add boiling water to gelatin mixture and stir until clear. Pour gelatin into ham mixture. Turn processor on long enough to mix.
6. Pour into loaf pan. Refrigerate 12 hours. Cut in slices and serve cold.

Nutrient Content/Serving:

Calories:	178	
Protein:	12.7	grams
Carbohydrate:	0.6	gram
Fat:	13.9	grams
Fiber:	2.2	grams

Equals two meat exchanges.

Garnish: Toast strips as tolerated.

Recipe Name: Chicken Casserole
Portion Size: 6 oz.
Servings: 8

Adapted to ___ Servings

6 1/2	oz.	Chicken Chunks (cooked)
3 1/4	oz.	Processed Cheese, grated
4	oz.	Cream of Chicken Soup, condensed
1/2	cup	Peas, drained
2	cups	Whole Milk
1	cup	Buttermilk Baking Mix
4	ea.	Eggs
1/4	tsp.	Salt
	dash	Nutmeg

1. Grease a half-size steam table pan, 2 1/2 inches deep.
2. Measure first 4 ingredients into food processor. Process until smooth, about 3 minutes.
3. Add remaining ingredients. Process until well mixed and smooth.
4. Pour into greased steam table pan.
5. Bake 30-35 minutes or until knife inserted in center comes out clean.
6. Place on hot food table. Serve with #6 disher or 6 oz. spoodle.

Nutrient Content/Serving:

Calories:	252	
Protein:	14.65	grams
Carbohydrate:	15.7	grams
Fat:	14.5	grams
Fiber:	0.65	gram

Equals two meat exchanges, one starch exchange, and one fat exchange.

Garnish: Parsley on side; cheese sauce if allowed.

Recipe Name: Tuna Salad
Portion Size: 1/8 loaf
Servings: 8

Adapted to ___ Servings

6 1/2	oz.	Tuna, drained
1/4	tsp.	Granulated Celery Seed
1/4	tsp.	Thyme
1/4	tsp.	Marjoram
2	ea.	Hard Cooked Eggs
1	cup	Diced Celery
1/4	tsp.	White Pepper
1/4	cup	Pickle Relish, Sweet (optional)
1/3	cup	Mayonnaise
3	env.	Unflavored Gelatin (1 env. = 1 Tbsp.)
1/4	cup	Cold Water
1/2	cup	Boiling Water

1. Pour cold water over gelatin and set aside to soak.
2. Measure first 8 ingredients into food processor. Process until smooth consistency.
3. Add mayonnaise and mix.
4. Add boiling water to soaked gelatin and stir until clear.
5. Add gelatin to mixture in food processor.
6. Pour into 9 × 4 1/2 × 4-inch loaf pan.
7. Refrigerate for 12 hours. Slice and serve cold.

Nutrient Content/Serving:

Calories:	150	
Protein:	10.2	grams
Carbohydrate:	3.5	grams
Fat:	10.6	grams
Fiber:	0.34	gram

Equals one meat and one fat exchange when pickle relish omitted.

Garnish: Hard cooked egg slices (add one additional meat exchange per egg).

Recipe Name: Turkey Pot Pie
Portion Size: 3/4 cup
Servings: 8

Adapted to ___ Servings

2	cups	Cooked and Cubed Turkey
1/2	cup	Mushrooms, drained
2	Tbsp.	Granulated Onion
1/2	tsp.	Salt
1	cup	Shredded Swiss Cheese
1 1/2	cups	Whole Milk
3/4	cup	Buttermilk Baking Mix
3	ea.	Eggs

1. Grease casserole or pie pan.
2. Heat conventional oven to 400° F or convection oven to 300° F .
3. Measure all ingredients and place in blender or food processor.
4. Process for 3 minutes or until a very smooth consistency.
5. Pour into casserole or pie pan and bake 30-35 minutes or until knife inserted in middle comes out clean.
6. Cool 5 minutes. Serve with a #6 disher or a 6 oz. spoodle.

Nutrient Content/Serving:

Calories:	248	
Protein:	18.7	grams
Carbohydrate:	10	grams
Fat:	14.8	grams
Fiber:	0.20	gram

Equals three meat exchanges and almost two vegetable exchanges.

Garnish: Celery seeds; gravy if allowed.

VEGETABLES AND FRUITS

Recipe Name: Pureed Vegetable
Portion Size: 1/4 cup
Servings: 10

Adapted to ___ Servings

1 1/4	cup	Cooked Vegetable
1/4	cup	Powdered Milk (no water added)
1/2	cup	Vegetable Juice or Broth (heated)
1/2	cup	Instant Potato (no water added)

1. Blend vegetable and powdered milk 3-4 minutes, gradually adding enough vegetable juice to achieve uniform texture and smooth consistency. Some vegetables with high water content must be drained well and will not need any broth.
2. Check with scoop for thickness. Add enough instant potato to hold vegetable together to scoop.
3. Serve with #16 scoop. A 1 oz. ladle of gravy may be served over vegetable.

Nutrient Content/Serving:			*With Gravy:*	
Calories:	14		59	
Protein:	1	gram	1	gram
Carbohydrate:	2.5	grams	2.5	grams
Fat:	0		5	grams
Fiber:	0.92	gram	0.92	gram

Equals one vegetable exchange if 1/2 cup is served. With sauce or gravy, add one fat exchange.

Garnish: Spices as appropriate; cheese sauce if permitted (see Sauces section for recipe).

Recipe Name: Spicy Carrot Puree
Portion Size: 1/2 cup
Servings: 5

Adapted to ___ Servings

1/2	cup	Water
4	cups	Sliced Carrots (about 8 medium)
1/2	cup	Ground Oat Flour
1/4	cup	Skim Milk
2	Tbsp.	Brown Sugar Substitute or Packed Light Brown Sugar
1	Tbsp.	Vegetable Oil or Margarine
1/2	tsp.	Salt (optional)
1/2	tsp.	Cinnamon
1/8	tsp.	Nutmeg

1. In large saucepan, bring water to a boil. Add carrots; cover. Simmer until tender, about 10 to 12 minutes.
2. Transfer carrots with water to food processor or blender; process until smooth.
3. Stir in remaining ingredients. At this point, puree may be placed in 1 1/2 qt. casserole. Cover and refrigerate until ready to use. Bake at 375° F about 30 minutes or until heated through.
4. Serve immediately.

Nutrient Content/Serving:

	With Artificial Sweetener:		*With Brown Sugar:*	
Calories:	91		112	
Protein:	2.7	grams	2.7	grams
Carbohydrate:	15.6	grams	20.9	grams
Fat:	2	grams	2	grams
Fiber:	6	grams	6	grams

Equals one starch exchange with the artificial brown sugar.

Garnish: Cinnamon, nutmeg, or brown sugar.

Note: In Spicy Carrot Puree, cinnamon and nutmeg enhance flavor without salt. Skim milk and a brown sugar substitute keep calories in check while oats and carrots provide 6 grams of fiber per serving. Try Spicy Carrot Puree as a satisfying side dish to prepare ahead.

Recipe Name: Mashed Sweet Potatoes
Portion Size: 2 oz.
Servings: 40

Adapted to ___ Servings

4	lbs.	Sweet Potatoes
6	oz.	Whole Milk, hot
4	oz.	Butter, soft
1/4	tsp.	Cinnamon, ground
as needed		Salt
1	tsp.	Grated Lemon Peel

1. Cook and peel sweet potatoes.
2. While potatoes are still warm, combine ingredients in mixer. Mix well.
3. Serve hot from steam table.

Nutrient Content/Serving:

Calories:	71	
Protein:	0.85	gram
Carbohydrate:	11	grams
Fat:	2.6	grams
Fiber:	1.6	grams

Equals one starch exchange.

Garnish: Dot with butter ball.

Hint: May be formed into patties for special occasions.

Recipe Name: Sweet Potatoes with Orange
Portion Size: 1/2 cup
Servings: 6

Adapted to ___ Servings

2	cups	Cooked Sweet Potatoes
2	Tbsp.	Margarine
1/2	ea.	Orange with Peeling (cut in quarters)
1/4	cup	Brown Sugar
2	ea.	Eggs

1. Puree first 4 ingredients in food processor until a very smooth consistency.
2. Add eggs and process for 30 seconds.
3. Pour into 1 quart greased baking pan.
4. Bake for 30 minutes at 350° F.
5. Serve hot, using #8 disher or 4 oz. spoodle.

Nutrient Content/Serving:

Calories:	190	
Protein:	4.1	grams
Carbohydrate:	30	grams
Fat:	6	grams
Fiber:	2.2	grams

Garnish: Melt marshmallows over top.

Recipe Name: Corn Souffle
Portion Size: 3/4 cup
Servings: 4

Adapted to ___ Servings

12	oz.	Yellow Corn Niblets
1/4	cup	Sugar
2	ea.	Eggs
1/4	tsp.	White Pepper (optional)

1. Drain corn, save liquid.
2. Place corn and sugar in food processor or blender; puree until smooth.
3. Add eggs and pepper. Turn food processor on and off for 30 seconds (overmixing will make mixture foam). Add reserved corn liquid (3-4 Tbsp.) one tablespoonful at a time until meringue-like consistency is reached.
4. Pour into 4 custard cups that have been sprayed with vegetable oil.
5. Place custard cups into a pan to which 1 inch water has been added.
6. Bake for about 30 minutes or until set (350° F for conventional oven or 275° F for convection oven).
7. Serve warm.

Nutrient Content/Serving:

Calories:	171	
Protein:	5.4	grams
Carbohydrate:	30	grams
Fat:	3.3	grams
Fiber:	2.82	grams

Garnish: Dash of pepper (even white) if permitted.

Hint: This dish will hold its shape much better in the individual molds.

Recipe Name: Honey Cinnamon Carrots
Portion Size: 1/2 cup
Servings: 6

Adapted to ___ Servings

4	cups	Carrots, sliced, fresh/frozen
1/2	cup	Chicken Broth
1/2	tsp.	Salt
1 1/2	tsp.	Cinnamon, ground
1/4	cup	Honey
2	Tbsp.	Margarine

1. Place carrots in boiling broth and steam until tender, about 20 minutes. (Frozen carrots need much less liquid.)
2. Remove from heat and drain.
3. Place in food processor and puree until smooth consistency.
4. Add remaining ingredients. Puree.
5. If not the consistency of mashed potatoes, add small amount of instant mashed potato powder.
6. Serve hot, using #8 disher or 4 oz. spoodle.

Nutrient Content/Serving:

Calories:	116	
Protein:	1	gram
Carbohydrate:	19	grams
Fat:	4	grams
Fiber:	3.2	grams

Omit honey and margarine to equal one vegetable exchange.

Garnish: Minced parsley.

Hint: Parsnips may be substituted for carrots in any proportion to make a vegetable medley.

Recipe Name: Congealed Watermelon
Portion Size: 1/4 cup
Servings: 32

Adapted to ___ Servings

2	qts.	Pureed Watermelon
3/4	cup	Cold Water
6	env.	Unflavored Gelatin (1 env. = 1 Tbsp.)

1. Remove seeds and puree watermelon to make 2 quarts.
2. Put 3/4 cup cold water in small saucepan. Add gelatin and allow to soak for 1 minute.
3. Heat gelatin mixture over low heat until clear.
4. Add to pureed watermelon and stir until mixed.
5. Pour into individual molds or 2 1/2 quart pan. Refrigerate 12 hours or overnight.

Nutrient Content/Serving:

Calories:	15	
Protein:	1.2	grams
Carbohydrate:	2.5	grams
Fat:	0	
Fiber:	0.4	gram

Equals one "free" exchange.

Garnish: Assorted gelatin cubes.

Recipe Name: Pureed Fresh Melon
Portion Size: 1/2 cup
Servings: 12

Adapted to ___ Servings

3	lbs.	Peeled and Cubed Cantaloupe or Honeydew
2	Tbsp.	Lemon or Lime Juice
1 cup + 2 1/2	Tbsp.	Instant Food Thickener

1. Place melon cubes in food processor bowl; blend until smooth.
2. Gradually add lemon or lime juice and blend until mixed thoroughly.
3. Add instant food thickener and blend well. Scrape bowl and blade and blend again until a smooth consistency.
4. Wait 30 seconds, then dish into bowls. Chill.

Nutrient Content/Serving:

Calories:	40	
Protein:	0.3	gram
Carbohydrate:	9.4	grams
Fat:	trace	
Fiber:	4.9	grams

Equals one fruit exchange with the garnish.

Garnish: #40 scoop of sherbet.

Recipe Name: Quick Applesauce
Portion Size: 1/2 cup
Servings: 4

Adapted to ___ Servings

12	ea.	Apples, medium-size (cored, peeled, and chopped)
1/2	cup	Water
1	tsp.	Cinnamon

1. Combine apples and water in 6 quart saucepan. Cook over medium heat until apples are soft (15-20 minutes).
2. Mash apples with a heavy spoon, or for a smoother consistency, blend in a food processor. Serve warm.

Nutrient Content/Serving:

Calories:	60	
Protein:	1	gram
Carbohydrate:	14	grams
Fat:	0	
Fiber:	2.6	grams

Equals one fruit exchange.

Garnish: Cinnamon.

Hint: Transparent apples are wonderful for sauce. These small, green apples are available around July 4. The harvest time varies but lasts for approximately two weeks. They are nicknamed the "June" apple. Due to their small size, fifteen are recommended for this recipe. (Some cooks are using applesauce in meat loaf and even BBQ sauce!)

SALADS

Recipe Name: Carrot-Raisin Salad
Portion Size: 1/2 cup
Servings: 12

Adapted to ___ Servings

2	lb.	Carrots, peeled and cut in 1 1/2-inch lengths
1	cup	Mayonnaise or Salad Dressing
1	tsp.	Salt
1	Tbsp.	Sugar
1 1/2	cup	Raisins
1/2	cup	Boiling Water

1. Puree carrots in food processor until a very fine consistency.
2. Mix in mayonnaise or salad dressing, salt, and sugar. Set aside.
3. In food processor, blend raisins for 1 minute.
4. Add boiling water; continue processing until a fine, smooth consistency.
5. To serve: place #8 scoop of carrot mixture into a 6 oz. fruit dish. Top with 2 tsp. (or #100 scoop) of raisin mixture. Serve cold.

Nutrient Content/Serving:

Calories:	230	
Protein:	1.4	grams
Carbohydrate:	23.1	grams
Fat:	14.7	grams
Fiber:	3.1	grams

Garnish: Raisin topping as described above.

Recipe Name: Diabetic Lime-Applesauce Gelatin
Portion Size: 3 × 3-inch square
Servings: 48

Adapted to ___ Servings

3 1/2	oz.	Diet Lime Gelatin Powder (dry)
11 1/2	cups	Applesauce
2 1/2	qts.	Diet Ginger Ale

1. In saucepan, combine gelatin and applesauce.
2. Cook and stir until gelatin dissolves.
3. Cool to room temperature.
4. Gently stir in diet ginger ale.
5. Pour into two 12 × 18 × 2-inch pans. Chill until firm.
6. Cut into squares. Refrigerate until serving time.

Nutrient Content/Serving:

Calories:	44	
Protein:	1.8	grams
Carbohydrate:	8.5	grams
Fat:	0.3	gram
Fiber:	1.2	grams

Equals 1/2 bread exchange.

Garnish: Dollop of applesauce on top.

Recipe Name: Mashed Banana Salad
Portion Size: 1/2 cup
Servings: 10

Adapted to ___ Servings

1/2	cup	Crushed Pineapple
2	ea.	Bananas, medium size
3	oz.	Cream Cheese
1	Tbsp.	Salad Dressing or Mayonnaise
1/4	cup	Peanut Butter
6	oz.	Lime Gelatin Powder (dry)
1 1/2	cups	Boiling Water
1 1/2	cups	Ice Cold Water (fill pint measure to 1 1/2 cup mark with ice cubes. Pour water over cubes to same mark.)

1. Measure first 5 ingredients into food processor or blender bowl.
2. Process until smooth. Set in refrigerator while preparing gelatin.
3. Spray ten 1/2 cup gelatin molds or two 6 × 3 1/2 × 2 1/2-inch pans with vegetable oil.
4. Dissolve gelatin in boiling water. Stir until well dissolved.
5. Stir in ice water. Allow to set until consistency of egg white. Stir in the blended ingredients.
6. Pour into gelatin molds or pans. Chill until set.
7. Serve over bed of bibb lettuce.

Nutrient Content/Serving:

Calories:	179	
Protein:	4.2	grams
Carbohydrate:	23.5	grams
Fat:	7.6	grams
Fiber:	0.7	gram

Garnish: Bed of lettuce leaves as described above.

Recipe Name: Pear Cottage Salad
Portion Size: 1/2 cup
Servings: 8

Adapted to ___ Servings

1	cup	Cottage Cheese
6	ea.	Pear Halves, drained; reserve juice
4	ea.	Lettuce Leaves
1/4	cup	Mayonnaise
1	env.	Unflavored Gelatin (1 env. = 1 Tbsp.)
2	Tbsp.	Cold Water
6	oz.	Cherry Gelatin Powder (dry)
1 1/2	cups	Boiling Water (use pear juice with water added)

1. Spray eight 1/2 cup gelatin molds with vegetable oil.
2. Puree cottage cheese in food processor until smooth.
3. Add pear halves and puree until smooth.
4. Add lettuce and puree until smooth.
5. Add mayonnaise and mix.
6. Soak unflavored gelatin in cold water.
7. Mix cherry gelatin with boiling water until dissolved. Add soaked unflavored gelatin. Stir until dissolved.
8. Add gelatin mixture to pureed cottage cheese mixture.
9. Pour into gelatin molds and allow to chill for 12 hours.
10. Unmold and serve on mustard greens, curly kale, or spinach leaves.

Nutrient Content/Serving:

Calories:	210	
Protein:	6.6	grams
Carbohydrate:	30.8	grams
Fat:	6.8	grams
Fiber:	0.75	gram

Garnish: Bed of greens as described above.

Recipe Name: Canned Fruit Aspic
Portion Size: 3 oz.
Servings: 10

Adapted to ___ Servings

5	cups	Canned Fruit, unsweetened; drained
1 1/2	env.	Unflavored Gelatin (1 env. = 1 Tbsp.)
1/8	cup	Cold Fruit Juice
1/4	cup	Boiling Water

1. Blend fruit in food processor for 1-2 minutes, or until uniform texture and consistency are achieved.
2. Spray with vegetable oil ten 1/2 cup gelatin molds or two 6 × 3 1/2 × 2 1/2-inch pans.
3. Sprinkle gelatin over cold fruit juice. Let stand until moistened.
4. Add boiling water and stir until gelatin is dissolved.
5. Add gelatin mixture to pureed fruit and blend for 1 minute or until a smooth consistency is achieved.
6. Pour into gelatin molds or pans. Refrigerate overnight.
7. When ready to serve, shake molds to loosen gelatin and turn out onto serving plate, or cut pans into 10 servings.
8. Serve on leaf lettuce.

Nutrient Content/Serving:

Calories:	54	
Protein:	1.4	grams
Carbohydrate:	12.2	grams
Fat:	trace	
Fiber:	1.5	grams

Equals one fruit exchange if unsweetened fruit used.

Garnish: Lettuce leaf as above, and with mayonnaise or sour cream beside aspic.

DESSERTS

Recipe Name: Peach Fluff
Portion Size: 1/2 cup
Servings: 6

Adapted to ___ Servings

4	ea.	Peaches
3	cups	Whipped Topping
2	Tbsp.	Powdered Sugar

1. Peel the peaches and mash them though a sieve.
2. Whip the topping and sugar together.
3. Stir in peaches and pile in sherbet glasses.
4. Chill.

Nutrient Content/Serving:

Calories:	166	
Protein:	1.4	grams
Carbohydrate:	8.5	grams
Fat:	14.1	grams
Fiber:	1.2	grams

Garnish: Cinnamon.

Recipe Name: Chocolate Mousse
Portion Size: 3 × 3-inch square
Servings: 12

Adapted to ___ Servings

1/2	gal.	Chocolate Ice Cream
1	cup	Graham Cracker Crumbs
1	cup	Saltine Cracker Crumbs
1/2	cup	Melted Butter
2	pkgs.	Instant Chocolate Pudding (4 1/2 oz.)
1 1/2	cups	Whole Milk
1	cup	Light Whipping Cream, whipped

1. Thaw ice cream.
2. Mix cracker crumbs with melted butter and ice cream.
3. Mix pudding according to package directions using only 1 1/2 cups milk.
4. Beat ice cream mixture with pudding. Pour into greased 12 × 9 × 2-inch pan.
5. Chill. Spread with whipped cream topping.

Nutrient Content/Serving:

Calories:	423	
Protein:	7.11	grams
Carbohydrate:	49.1	grams
Fat:	22	grams
Fiber:	0.8	gram

Garnish: Crisp topping (see recipe, page 77) for extra fiber, as tolerated.

Recipe Name: Strawberry Sponge
Portion Size: 1 slice from mold or 1/4 cup
Servings: 6

Adapted to ___ Servings

1	env.	Unflavored Gelatin (1 Tbsp.)
1	cup	Strawberries
1/4	cup	Cold Water
1/2	cup	Hot Water
1	Tbsp.	Lemon Juice
1	cup	Sugar
1/2	cup	Cream or Evaporated Milk, whipped
	pinch	Salt
2	ea.	Egg Whites, beaten stiff

1. Crush strawberries. Add sugar and allow to stand 30 minutes.
2. Soften gelatin in cold water. Add salt and hot water; stir until dissolved. Add gelatin mixure and lemon juice to the strawberries.
3. Cool. When it begins to thicken, beat well. Fold in the whipped cream and stiffly beaten egg whites (these must be beaten separately). Turn into a mold or individual 2 oz. custard cups. Chill.

Nutrient Content/Serving:

Calories:	220	
Protein:	3.9	grams
Carbohydrate:	47.3	grams
Fat:	1.7	grams
Fiber:	0.5	gram

Garnish: Dry gelatin.

Recipe Name: Applesauce Refrigerator Cake
Portion Size: 2 × 4-inch rectangle
Servings: 8

Adapted to ___ Servings

1	can	Sweetened Condensed Milk (15 oz.)
1/2	cup	Lemon Juice
2	cups	Applesauce
5 3/4	oz.	Vanilla Wafers (approximately 48), divided
1 1/2	tsp.	Cinnamon, divided
	dollop	Whipped Cream

1. Line an 8 × 8-inch pan with plastic film.
2. Place a layer of one-third of the vanilla wafers on bottom of pan.
3. In food processor, mix condensed milk and lemon juice. Process until thickened. Spread over layer of vanilla wafers.
4. Sprinkle half of the cinnamon on condensed milk mixture.
5. Place another layer of vanilla wafers on top. Spread 2 cups of applesauce on top of vanilla wafers.
6. Sprinkle remaining cinnamon on top of applesauce.
7. Layer remaining vanilla wafers on top. Cover and refrigerate overnight (12 hours or more).
8. At serving time, cut into 2 × 4-inch pieces. Top with whipped cream and garnish.

Nutrient Content/Serving:

Calories:	350	
Protein:	6.2	grams
Carbohydrate:	59	grams
Fat:	9.6	grams
Fiber:	0.8	gram

Garnish: Pureed maraschino cherries or red gelatin cube.

Recipe Name: Peanut Butter Pudding
Portion Size: 1/2 cup
Servings: 24

Adapted to ___ Servings

2	cup	Honey
1	cup	Peanut Butter
3/4	cup	Cornstarch
1/4	Tbsp.	Salt
2	qts.	Whole Milk (scalded)
1	Tbsp.	Vanilla Extract

1. Blend together first four ingredients until smooth.
2. Whip into milk over low heat. Cook until thickened, stirring constantly.
3. Stir in vanilla. Cool and serve with garnish.

Nutrient Content/Serving:

Calories:	228	
Protein:	5.6	grams
Carbohydrate:	32.8	grams
Fat:	8.3	grams
Fiber:	0.67	gram

Garnish: Whipped topping.

Recipe Name: Easy Peach Torte
Portion Size: 1/8 pie
Servings: 8

Adapted to ___ Servings

3	ea.	Egg Whites
1	cup	Sugar
1/2	tsp.	Baking Powder
12	ea.	Graham Cracker Squares, crushed
10	ea.	Peaches, sliced
1/2	pt.	Whipping Cream, whipped
1/2	tsp.	Vanilla Extract

1. Beat egg whites. Gradually add sugar and baking powder. Fold in the crushed graham crackers.
2. Pour into well-greased 9-inch pie pan and bake at 350° F for 30 minutes.
3. Puree sliced peaches in their own juice (or strawberries; see Hint); set aside.
4. Whip the cream with the vanilla extract; set aside.
5. When cake has cooled, spread pureed fruit over top, then spread with whipped cream. Refrigerate until ready to serve.

Nutrient Content/Serving:

Calories:	191	
Protein:	2.5	grams
Carbohydrate:	32.7	grams
Fat:	5.6	grams
Fiber:	0.96	gram

Garnish: Graham cracker crumbs with cinnamon and sugar.

Hint: Substitute 8 cups sliced strawberries, pureed, for peaches.

Recipe Name: Pumpkin Delight
Portion Size: 1/2 cup
Servings: 3

Adapted to ___ Servings

3/4	cup	Canned Pumpkin
3/4	cup	Whole Milk
1	ea.	Egg
1	Tbsp.	Artificial Sweetener (or 1/4 cup Sugar)
1/4	tsp.	Vanilla
1/2	tsp.	Cinnamon
1/4	tsp.	Nutmeg
1/8	tsp.	Salt

1. Beat egg; add milk, vanilla, sweetener, salt, cinnamon, and nutmeg; mix.
2. Pour into large bowl with canned pumpkin and mix well.
3. Ladle into 3 custard cups; place cups in a pan with one inch deep hot water. Oven poach at 350° F for 50 to 60 minutes, or until a knife inserted near the center comes out clean.
4. Serve chilled.

Nutrient Content/Serving:

	With Artificial Sweetener:		*With Sugar/Serving:*	
Calories:	90		218	
Protein:	5	grams	5	grams
Carbohydrate:	7.8	grams	39.8	grams
Fat:	4.3	grams	4.3	grams
Fiber:	1.4	grams	1.4	grams

Equals 1/2 whole milk exchange with artificial sweetner.

Garnish: Whipped cream cheese.

Recipe Name: Bavarian Cream
Portion Size: 1/2 cup
Servings: 16

Adapted to ___ Servings

2	Tbsp.	Unflavored Gelatin	1	tsp.	Vanilla Extract
1/2	cup	Whole Milk	5	ea.	Egg Whites
1 2/3	cups	Whole Milk	1/8	tsp.	Salt
1	cup	Sugar	1/4	tsp.	Cream of Tartar
8	ea.	Egg Yolks	1/2	cup	Chilled Heavy Cream

1. Sprinkle gelatin over 1/2 cup of milk to soften; heat 1 2/3 cup milk.
2. Separate eggs, reserving 5 whites. Gradually whip sugar into the egg yolks, beating several minutes until pale and thick.
3. Then, by dribbles, beat the hot milk into the yolks and sugar, thus heating the yolks slowly to prevent curdling. Set on moderate heat and stir slowly with wooden spoon. Remove from heat instantly if mixture starts to lump, then continue over heat until it begins to thicken like a custard.
4. Remove from heat and stir for 1 minute. Stir the softened gelatin mixture over heat until dissolved. Beat in the hot custard. Stir in vanilla.
5. Beat the egg whites until they begin to foam. Add the salt and cream of tartar; continue beating until the peaks begin to stiffen. Fold into the hot custard mixture.
6. Beat cream in large bowl until it doubles in volume and holds its shape softly. Set aside.
7. Set custard-gelatin bowl in a bowl of ice and fold with spatula every few minutes to cool and thicken until almost set (8 to 10 minutes). Fold cream into custard.
8. Pour mixture into a wet gelatin mold (a mold dipped in chilled water). Refrigerate until ready to serve.

Nutrient Content/Serving:

Calories:	134	
Protein:	4.5	grams
Fat:	6.5	grams
Carbohydrates:	14.4	grams
Fiber:	trace	

Garnish: Dust with cocoa or glaze with caramel sauce.

Recipe Name: Chocolate Meringue Pie
Portion Size: 1/8 pie
Servings: 8

Adapted to ___ Servings

Filling:	4	ea.	Egg Yolks
	1 1/2	cups	Sugar
	3	Tbsp.	Cocoa
	1/2	cup	Buttermilk Baking Mix
	1	cup	Whole Milk
	1/4	cup	Margarine, melted
	1 1/2	tsp.	Vanilla Extract
Meringue:	4	ea.	Egg Whites (room temperature)
	1/4	tsp.	Cream of Tartar
	1/4	tsp.	Salt
	1/2	cup	Sugar

Filling:
1. Grease a 9 1/2 × 1 1/4-inch pie pan and set aside.
2. Separate 4 eggs and save egg whites for meringue.
3. Measure all filling ingredients into food processor. Process until smooth, about 1 1/2 minutes. Pour into greased pie pan.
4. Bake at 350° F in conventional oven or 275° F in convection oven for about 40 minutes, or until set in middle.

Meringue:
1. Place egg whites, cream of tartar, and salt in 2-3 quart bowl or mixer bowl. Begin beating with wire beater or mix at high speed. As eggs become frothy, begin adding sugar while beating. Continue to beat until peaks form.
2. Spread meringue on top of baked pie while still hot.
3. Bake at 350° F on middle rack of conventional oven (or 275° F in convection oven, on top or bottom rack) until browned, about 10-12 minutes. Cool one hour in refrigerator before serving.

Nutrient Content/Serving:

Calories:	352				
Protein:	5.3	grams	Fat:	11.3	grams
Carbohydrate:	57.2	grams	Fiber:	0.0	grams

Garnish: Thinly shaved chocolate.

SAUCES

Recipe Name: Pineapple Sauce
Portion Size: 2 oz.
Servings: 8

Adapted to ___ Servings

1/2	cup	Brown Sugar
1	tsp.	Cornstarch
1	cup	Water
1	cup	Pineapple Juice, unsweetened

1. Mix brown sugar and cornstarch.
2. Bring water and pineapple juice to a boil.
3. With wire whisk, stir in sugar/cornstarch mixture.
4. Stir until clear and thickened.
5. Serve hot, using 2 oz. ladle, over hot pureed ham.

Nutrient Content/Serving:

Calories:	54	
Protein:	0.13	gram
Carbohydrate:	13.3	grams
Fat:	0.04	gram
Fiber:	0	

Recipe Name: Raspberry Sauce
Portion Size: 1 Tbsp.
Servings: 16

Adapted to ___ Servings

2	cups	Fresh Raspberries
2	Tbsp.	Sugar

1. Combine ingredients in 2 qt. microwaveable casserole. Microwave at high power for 4-6 minutes, or until boiling rapidly, stirring halfway through cooking time.
2. Let cool slightly.
3. Stir briskly to break up the berries somewhat, or puree in blender or food processor for a more syrup-like sauce. Makes 1 cup.

Nutrient Content/Serving:

Calories:	16	
Protein:	0	
Carbohydrate:	4	grams
Fat:	0	
Fiber:	2.2	grams

Serving Suggestion: As topping for cake, sherbet, ice cream, and chicken, or use as a liner for custards and puddings. Peaches may be substituted.

Recipe Name: White Sauce Proportions
Portion Size: 8 oz.
Servings: 1

Type	Milk	Flour	Fat	Salt	Usage
Thin	8 oz.	1 T.	1 T.	1/2 t.	Soups*
Medium	8 oz.	2 T.	2 T.	1/2 t.	Sauces**
Thick	8 oz.	3 T.	2 T.	1/2 t.	Souffles***

1. Melt the fat in heavy pan or in the top pan of a double boiler.
2. Add flour to make a roux. Stir to a paste.
3. Stir in milk slowly. Continue stirring.
4. Add salt. Cook until thickened, stirring occasionally.

Nutrient Content/Serving:

	Thin	Medium	Thick	
Calories:	303	378	495	
Protein:	9.8	9.8	10	grams
Carbohydrate:	18	22	27.5	grams
Fat:	21.8	39	39	grams
Fiber:	0	0	0	

Variations:

*For a delicious cheese soup, blend 3/4 to 1 cup grated cheddar cheese into 8 oz. thin white sauce. Serve hot.

**Simply add 3/4 cup grated cheddar cheese to 8 oz. medium white sauce for a wonderful cheese sauce. Great over vegetables or macaroni.

***Preheat oven to 325° F. To 8 oz. thick white sauce, add 1 cup grated cheddar cheese. Heat thoroughly. Separate 3 eggs and beat yolks until light. Stir yolks into white sauce. Beat egg whites until stiff. Fold egg whites quickly into white sauce. Transfer to a greased baking dish using rubber spatula. Bake 1 hour, or until a knife inserted at the center comes out clean. Serve immediately.

Garnish for all Variations: Paprika or, if tolerated, black pepper.

SUPPLEMENTS

Recipe Name: Eggnog
Portion Size: 1 cup
Servings: 5

Adapted to ___ Servings

2	ea.	Egg Yolks
2	cups	Evaporated Milk, divided
1/4	tsp.	Salt
2	cups	Water
2	ea.	Egg Whites
1-2	tsp.	Vanilla Extract
as needed		Brandy or Rum Flavoring, to taste
6	pkts.	Artificial Sweetener (or 1/4 cup Sugar)

1. Separate eggs, reserving whites. Beat egg yolks with 1/4 cup of the evaporated milk. Stir in salt, water, and rest of the milk.
2. Cook over medium heat, stirring constantly, until the mixture coats the spoon. Cool.
3. Beat egg whites until they form soft peaks. Add to custard and mix thoroughly.
4. Add sweetener and flavorings as desired.
5. Serve warm or chilled.

Nutrient Content/Serving

	With Artificial Sweetener:		*With Sugar:*	
Calories:	125		138	calories
Protein:	8	grams	8	grams
Carbohydrate:	12	grams	15.2	grams
Fat:	5	grams	5	grams
Fiber:	0		0	

Equals one 2% milk exchange.

Garnish: Sprinkle with nutmeg before serving.

Recipe Name: Fortified Milk
Portion Size: 1 cup
Servings: 1

Adapted to ___ Servings

1/4	cup	Nonfat Dry Milk
1	cup	Whole Milk

Combine ingredients in blender. Serve chilled.

Nutrient Content/Serving:

Calories:	267	
Protein:	19.3	grams
Carbohydrate:	27.7	grams
Fat:	8.8	grams
Fiber:	0	

Recipe Name: Power Shake
Portion Size: 8 oz.
Servings: 4

Adapted to ___ Servings

2	cups	Whole Milk
1	cup	Vanilla Ice Cream
2	ea.	Bananas
6	Tbsp.	Vegetable Oil (1/4 cup + 2 Tbsp.)
2	oz.	Instant Breakfast Mix
1	cup	Eggnog Powder
10	Tbsp.	Egg Powder (1/2 cup + 2 Tbsp.)
1	cup	Nonfat Dry Milk

Blend all ingredients together. Refrigerate before serving.

Optional: Double the dietary fiber by adding 1/2 cup of processed bran.

Nutrient Content/Serving:			*With Bran:*	
Calories:	941		969	
Protein:	67.1	grams	68.4	grams
Carbohydrate:	108.7	grams	113.2	grams
Fat:	26.4	grams	27	grams
Fiber:	3	grams	6.2	grams

Garnish: Dollop of whipped topping.

BREADS/CEREAL AND TOPPINGS

Recipe Name: Super Cereal
Portion Size: 8 oz.
Servings: 24

Adapted to ___ Servings

1	qt.	Water
1	cup	Nonfat Dry Milk
6	cups	Evaporated Milk (four 12 oz. cans), divided
2 2/3	cups	Oatmeal
1 1/2	lbs.	Margarine
1	lb.	Brown Sugar
2	cups	Granulated Sugar
3	cups	Processed Bran (optional)

1. Combine water, nonfat dry milk, and 1 1/2 cups (one 12 oz. can) evaporated milk. Bring to a boil.
2. Pour in oatmeal and cook about 5 minutes.
3. Add margarine, sugars, and remainder of evaporated milk. Stir until creamy.
4. Add processed bran if desired, and stir well.

Nutrient Content/Serving:			*With Bran:*	
Calories:	491		506	
Protein:	8	grams	9	grams
Carbohydrate:	54	grams	56	grams
Fat:	27	grams	27.3	grams
Fiber:	1.28	grams	3	grams

Garnish: Brown sugar, jam, pureed fruit (applesauce), honey, or syrup.

Recipe Name: Banana Tea Bread
Portion Size: 1/2-inch
Servings: 32 (Cut into 16 1/2-inch pieces, then divide them in half)

Adapted to ___ Servings

2 1/2 cups	Flour
1 tsp.	Baking Powder
1/2 tsp.	Baking Soda
1/4 tsp.	Salt
1/2 cup	Prune Puree (See Notes) or Prepared Prune Butter
1 cup	Sugar
1/2 cup	Water
1 tsp.	Vanilla
1 1/4 cups	Very Ripened, Mashed Bananas (4 medium)
1/4 cup	Buttermilk (or mix scant tsp. vinegar or lemon juice with 1/4 c. low fat milk. Allow to stand 5 min.)

1. Preheat oven to 350 degrees Fahrenheit.
2. Sift flour, baking powder, baking soda, and salt together.
3. Cream prune puree with sugar, water, and vanilla.
4. Mix in mashed bananas.
5. Grease a 9 × 5-inch bread pan.
6. Add sifted ingredients and buttermilk alternately, beating well after each addition.
7. Pour into greased loaf pan.
8. Bake in center of oven for 1 hour, 15 minutes until pick inserted into center comes out clean.
9. Cool in pan and turn out onto rack for complete cooling. For best flavor, wrap and store overnight before serving.

Nutrient Content/Serving:

Calories:	77	
Protein:	1.5	grams
Carbohydrate:	18	grams
Fat:	0.2	grams
Fiber:	0.8	grams

Equals 1/2 bread and 1/2 fruit exchange.

Notes: Prune Butter Substitute: Combine 2/3 cup pitted prunes and 3 Tbsp. water in food processor. Pulse until prunes are pasty.

Recipe Name: Southern Corn Bread
Portion Size: 3 oz.
Servings: 20

Adapted to ___ Servings

6	cups	Whole Milk, divided
12	oz.	Yellow Cornmeal (coarsely, ground)
8	oz.	Butter, melted
4	tsp.	Baking Powder
1	Tbsp.	Salt
8	ea.	Eggs, lightly beaten

1. Preheat conventional oven to 350° F. Bring 4 cups of the milk to a boil in large saucepan over medium-high heat.
2. Gradually add cornmeal to milk. Stir rapidly with wire whisk to incorporate. Reduce heat to low. Cook 5-8 minutes, stirring constantly. Remove from heat. Cool to lukewarm (approximately 15 minutes).
3. Combine melted butter, remainder of milk, baking powder, and salt in small bowl. Add to cornmeal mixture. Stir well.
4. Add beaten eggs to cornmeal mixture. Stir well.
5. Pour into buttered 4 qt. baking pan. Bake 40-45 minutes, or until knife inserted in center comes out clean.

Nutrient Content/Serving:

Calories:	213	
Protein:	6.1	grams
Carbohydrate:	15.4	grams
Fat:	14.1	grams
Fiber:	1.6	grams

Garnish: Apple or herb butter.

Note: Soaked cornbread is more appropriate for the brittle diabetic on a pureed diet.

Recipe Name: Banana Frosting
Portion Size: 2 × 3-inch rectangle
Servings: 16

Adapted to ___ Servings

5	Tbsp.	Egg Whites from 3 Large Eggs
3	Tbsp.	Sugar
6 1/2	oz.	Banana Puree from 1 Large Banana

1. Beat egg whites until frothy. Gradually beat in sugar and banana puree to form stiff peaks.
2. Spread on cooled 8 × 12-inch cake.
3. Bake at 500° F until lightly browned.

Nutrient Content/Serving:

Calories:	27	
Protein:	0.6	gram
Carbohydrate:	6	grams
Fat:	trace	
Fiber:	0.22	gram

Garnish: Drizzle of chocolate sauce.

Recipe Name: Crisp Topping
Portion Size: 1 Tbsp.
Servings: 16

Adapted to ___ Servings

2	Tbsp.	Margarine
1	cup	Graham Cracker Crumbs
as desired		Garlic Powder or Parmesan Cheese

1. Melt margarine in a small pan over medium heat.
2. Add graham cracker crumbs, cook and stir 2-3 minutes.
3. Remove from heat and toss with garlic powder or Parmesan cheese. As the mixture cools, the cracker crumbs will become crispy.
4. Sprinkle on soups, vegetables, etc.

Nutrient Content/Serving:

Calories:	34	
Protein:	0.4	gram
Carbohydrate:	3.9	grams
Fat:	1.9	grams
Fiber:	1.05	grams

Note: High-fiber graham crackers are now available and may be substituted to achieve a higher dietary fiber content.

GLOSSARY OF FOOD TERMS

Aspic – A jelly made from meat stock that has been boiled down sufficiently to become firm when cold; also stock (fish, vegetable, fruit, or tomato juice) which has been thickened with gelatin.

Custard – A true one consists only of eggs, milk, sugar, and flavoring. Hint: for a variation, mix 1 1/2 cup bread cubes with every 1 3/4 cup milk before baking.

Dollop – 1/2 tsp.

Frappe – Sweetened fruit juice frozen to a mushy consistency.

Garnish – Any edible accent added to a dish to decorate one food with another.

Glaze – To coat with a thin sugar syrup that has been cooked to the crack stage, or with thin aspic or diluted fruit jelly.

Mousse – Sweetened and flavored whipped cream which may contain a small amount of gelatin.

Pinch – 1/8 tsp.

Quiche – An unsweetened, open-faced custard pie.

Roux – Equal parts of fat and flour that have been heated without browning. As a thickener, it may be used in white sauce before liquid (usually milk) is added.

Scald – To heat milk until small bubbles appear where milk comes in contact with sides of pan. Because milk scorches easily, it is safer to scald milk over hot water than over direct heat.

Souffle – Main dish or dessert mixture that may be lightened or raised with stiffly beaten egg whites. If a main dish, it is baked and must be served immediately after removal from the oven. It will hold in the oven when turned off for 10 minutes but once removed will hold just 20 minutes. If a dessert, it is served hot or cold.

Sponge – A beaten syrupy gelatin dispersion into which whipped egg white is folded.

Stabilizer – An agent that adds body (e.g., gelatin, which absorbs water and gives a smooth texture).

Strata – A combination of bread baked in an egg custard.

Torte – A rich, layered cake.

REFERENCES

Anderson, James et al. *Professional Guide to HCF Diets*. Lexington, KY: HCF Diabetes Research Foundation, 1981.

USDA. *Nutritive Value of American Foods in Common Units*. Washington, DC: Superintendent of Documents, U.S. Government Printing Office, Agriculture Handbook No. 456, 1981.

American Diabetes Association and American Dietetic Association. "Exchange Lists for Meal Planning." 1989.

CLINICAL TOPICS

I. Current Trends in Geriatric Nutrition

Current research in geriatric nutrition indicates that there is a dynamic relationship between diet and disease. While it has long been known that diet influences the course of diabetes, obesity, hypertension, and weight loss, more recently cancer and osteoporosis have been implicated as diet-related conditions. This evidence suggests that each of these conditions merits investigation.

The correlation between diet and diabetes management has been well-documented. Balanced daily meals planned according to the diabetic exchange system on a routine schedule with an evening nourishment provide a basis for caloric and blood sugar control. Planned timing of meals allows for proper balance between carbohydrate delivery and the amount of circulating insulin (a hormone) in the body.[1] Insulin is produced in the pancreas by the Islets of Langerhans beta cells. Generally, these cells decline in number with advancing age. Insulin can be thought of as the key that unlocks the door to our body cells, allowing carbohydrates such as glucose to enter and be oxidized to produce energy. If the insulin is not released into the bloodstream, the cells gain no energy. Subsequent symptoms such as polyuria (frequent urination), polydipsia (frequent thirst), polyphagia (frequent hunger), weight loss, fatigue, and increased susceptibility to infection may occur. Complicating this situation for the elderly are decreased insulin secretion, physical inactivity, decreased dietary intake, decreased lean body mass, and insulin antagonism.[2]

We have reviewed the implications of inadequate insulin. Poor diet intake and a sedentary activity level only partially explain the problems of oral intake and energy usage. Glucose is the major stimulus for insulin secretion, but amino acids, fatty acids, and ketones have a similar but more depressed effect. When there is a lack of glucose ingestion, the mobilization of protein and fat through the mechanism of other hormones consequently raises blood sugar levels, and additional insulin is needed for their absorption and metabolism. Exercise requires energy from body cells; therefore, lack of exercise prevents glucose absorption into cells and metabolism for energy production.

The gradual decline of lean body mass in the life cycle causes less tissue to be available to store ingested carbohydrates. This may lead to impaired disposal of glucose. Insulin antagonism, or tissue sensitivity to insulin, results in impairment of insulin's role as the key that unlocks the door to body cells, thus allowing glucose to pass from the extracellular fluids through the cell membranes into the cell. Peripheral tissues such as muscle and fat have been cited in research literature as major antagonistic cells. Lab tests used to determine serum glucose include fasting blood sugar, random glucose, two hour post-prandial, and glycosylated hemoglobin test. In the last test, the higher the glucose level and the longer the time period in which it is elevated, the more glycosylated hemoglobin will be formed.

From our previous discussion, it is evident that proper diet, medication, and exercise can control the symptoms of diabetes in the elderly. Principles of diet therapy in the treatment of diabetes concentrate on covering the peak (meeting the maximum level) of insulin activity and preventing hypoglycemia during sleep. At least one nourishment should be given as a bedtime snack to fuel the fire of the blood sugar balance mechanism while the diabetic rests. Since protein is digested more slowly than carbohydrate, the h.s. (hora somni, Latin for "before sleep") snack must contain adequate protein. A balance of carbohydrate (50 percent), protein (20 percent), and fat (30 percent) is advocated by the American Diabetes Association. The predisposition to atherosclerotic heart disease can be reduced if the amount of total fat and cholesterol content is kept to 30 percent and 225-300 mg. for men and women, respectively, as recommended by the American Heart Association. Soluble dietary fiber also predisposes lower levels of serum glucose and low-density lipoprotein cholesterol. Because dietary fiber–both soluble and insoluble–may be beneficial, the American Diabetes Association marks foods that contain three grams or more with a special leaf symbol in their "Exchange Lists for Meal Planning."[3] (Soluble dietary fiber may help lower serum glucose.)

Individualization of the diet, and assessment of compliance through the use of food diaries, are most important; a calories/carbohydrate distribution of 2/10, 1/10, 2/10, 1/10, 3/10, 1/10 occurring at breakfast, mid-morning snack, lunch, mid-afternoon snack, supper, and evening snack, respectively, recommends itself. However, if one snack at bedtime only is considered practical, a distribution of 3/7, 3/7, 3/7, 1/7 for breakfast, lunch, supper, and bedtime snack is acceptable.

Of course, patients using an insulin pump control the timing of insulin delivery and the dosage of insulin according to the time and size of each meal. A registered dietitian should be consulted to instruct the individual and/or family about the diet order. Caloric intake should be sufficient to maintain ideal body weight. Overweight diabetics must reduce their weight to avoid the risks of complications, but any exercise regimen should be approved by a physician. Mono- or polyunsaturated fatty acids and lower cholesterol content reduce the predisposition to the development of atherosclerotic disease.

When diabetic persons are ill and cannot consume a normal diet, a liquid diet may be followed with physician consent. Here are two examples that contain approximately 1200 kilocalories each to build upon:

	Clear Liquid Diet	*Full Liquid Diet*
8 a.m.	2 fruit juice exchanges 1/2 cup jello (sweetened) coffee or tea	1 fruit juice exchange 1/2 cup hot cereal 1/2 cup whole milk coffee or tea
10 a.m.	broth 1 fruit juice exchange	1/2 cup whole milk 2 saltine crackers
noon	broth 1 fruit juice exchange 1/2 cup jello (sweetened) coffee or tea	1/2 cup cream soup 1 fruit juice exchange 2 saltine crackers coffee or tea

2 p.m.	1 fruit juice exchange	2 graham crackers 1/2 cup whole milk
5 p.m.	broth 1 fruit juice exchange 1/2 cup jello (sweetened) coffee or tea	1/2 cup cream soup 2 saltine crackers custard or ice cream coffee or tea
8 p.m.	1 fruit juice exchange 1/2 cup jello (sweetened)	2 saltine crackers 1/2 cup whole milk

OTHER NUTRITION-RELATED CONDITIONS

If hypertension results from a preexisting medical condition such as diabetes, it is termed a secondary condition. About 10 percent of hypertensive individuals are in this category. Most cases are classified as essential, i.e., hypertension for which there is no known specific cause. Major risk factors include cardiac heart disease (especially congestive heart failure), peripheral vascular disease, and chronic renal failure. These conditions warrant a modified sodium level of about two grams, especially in cases of right heart failure. The U.S.D.A. recognizes excess sodium and obesity as factors that affect blood pressure or hypertension in some people.[4] It is difficult for any age group to comply with a low-sodium diet regimen, but it is perhaps more challenging for the elderly person. Our taste and smell sensations diminish by approximately 65 percent with age, and although this change increases the perceived need for salt and sugar use, the elderly should decrease their intake of these additives. Proper oral hygiene improves taste perception, and many seasonings (e.g., dry mustard, lemon, allspice, nutmeg, parsley, chili powder, horseradish, vinegar, and paprika) help to improve the palatability of a low-sodium diet. (A note to kitchen managers: spices may lose their flavor after six months; therefore, procure in usable quantities.)

Salt acts as a fluid retainer in our bodies and elevates blood volume and subsequent pressure. Blood pressure is defined as the pressure exerted by blood on the walls of blood vessels (especially arteries) expressed in terms of systolic/diastolic measures. Achievement of ideal body weight reduces stress on the heart as it pumps blood through arteries, veins, and fat tissues. Diet restrictions should include a reduction of cheese, cured meats, potato chips, pickles, sauerkraut, salted nuts, and sauces. If salt substitute (potassium chloride) is requested, a doctor's order becomes necessary.

The National Osteoporosis Foundation estimates that 25 million Americans have some degree of osteoporosis.[5] It affects one-third of post-menopausal women and a majority of individuals over age 70. (Clinical studies have demonstrated that a high calcium intake, which serves to defend against osteoporosis, also decreases blood pressure in some individuals.[6] The National Institute of Health in its fifth report on the "Detection, Evaluation, and Treatment of High Blood Pressure," advises adequate calcium, magnesium, and potassium intakes.) After menopause, the body's absorption of bone mass becomes more rapid than deposition and yields a net loss of bone. Clinical symptoms such as back pain, loss of height, or excessive curvature of the upper back (dowager's hump) may follow, due to compression of vertebrae. Bone fractures occur more frequently and loss of the alveolar ridge under the gums may lead to loosening of the teeth and gum disease.

Diet therapy alone does not satisfactorily treat osteoporosis. However, adequate intake of calcium throughout the life cycle is extremely important in promoting bone mineralization. Calcium absorption drops by one-third between the ages of 70 and 90 as compared with 20- to 59-year olds.[7] A diet high in calcium with adequate levels of Vitamin D and trace minerals fluoride and boron helps to maintain bone repair. In the elderly, reductions occur both in the synthesis of Vitamin D in the skin (sometimes due to low sun exposure) and in the formation of the active hormone 1,25-dihydroxyvitamin D_3 in the kidney.[8] Therefore, it is recommended that the diet include two eight-ounce glasses of whole, Vitamin D fortified milk, which supplies approximately 600 of the 800 mg. of calcium needed to meet the requirement for persons over 50 years old. Other foods rich in calcium include yogurt, cheese, collard, kale, and mustard or turnip greens. If further vitamin, mineral, or hormonal supplementation is required, the physician may prescribe it.

As the second leading cause of death in the U.S., cancer remains a complex disease. A link between diet (an environmental causal factor) and cancer seems evident in 33 percent of cases. Colon and rectal cancers are the second most frequently diagnosed diseases in the U.S., excluding some skin cancers. Research reveals that a high fat and protein and low fiber diet plays a contributory role in the development of colon cancer.[9,10] Dr. Zbigniew Walaszek, M.D., of the University of Texas, has even isolated a substance from vegetables that seems to interfere with tumor growth in lab animals. The compound glucarate alters hormone levels that regulate cell growth. It is found in cruciferous vegetables from the mustard family, whose members have cross-like, four-petaled flowers. Examples include broccoli, cabbage, brussels sprouts, cauliflower, and bean sprouts.[10]

Since cancer cachexia (wasting) accounts for almost half of malnutrition cases in older, institutionalized adults, let's focus on side effects of cancer therapy. Chemotherapy, immunotherapy, anti-neoplastic agents, and even radiation can cause anorexia, nausea, or vomiting, and an aversion to meats and sweet foods may develop. The following list offers some suggestions for nutritional support:

Appetite or Weight Loss: Attempt smaller, more frequent feedings rather than three large meals. Add calories to foods by using sauces, gravies, cream, margarine, mayonnaise, sugar, and jellies. Special nutrient-dense snacks and supplements can be provided between meals and/or in the evening. Example: Nonfat dry milk may be added to whole milk to increase calories, not bulk, and used in cooking and as a beverage. Cater to food preferences and special requests.

Nausea and Vomiting: Offer smaller, more frequent meals. Advise slow drinking of liquids. It may help to withhold liquids at mealtimes, providing them 30-60 minutes before or after meals. Rest is advised after eating, but the patient should not lay down flat. The head should be kept at least four inches higher than the feet at a 30° angle.

Altered Taste and Smell Perceptions: Try egg, cheese, and peanut butter as meat substitutes. Experiment with extra seasoning such as salt, pepper, sugar, lemon juice, vinegar, mustard, ketchup, etc. If foods taste too sweet, add a pinch of salt. If meat is offered, serve chilled.

Dry Mouth or Sore Throat: Cold foods such as ice cream, pudding, and jello may be soothing. Combine dry foods with sauces, gravy, broth, juices, or other liquids. These may be poured over the food item or used as a dip. Avoid rough, raw, acidic, or spicy foods. Oral hygiene is very important. Keep mouth clean by brushing teeth and rinsing with prescribed mixture. Lozenges and hard candies may be helpful.

Viscous Mucous Production in Mouth: Suggest hot tea with lemon or other hot beverages and very cold beverages such as popsicles, Kool-aid, and canned sodas. Lemon drops or sour balls may help as well.

Diarrhea: Replace fluid loss with clear juices high in potassium. Be sure to rule out lactose intolerance or other food allergies. Attempt small, frequent meals. Suggest pectin-containing foods such as bananas and applesauce. Try tapioca, peanut butter, and cooked rice. Avoid high-fiber foods, high-fat foods, or gas-forming foods. Go easy on high-caffeine and spicy foods if bothersome.

Constipation: Rule out obstruction. Suggest increasing dietary fiber (bran, raw or cooked fruits and vegetables) and fluid intake, especially water, prune juice, or other fruit juices. Avoid high-fat foods. Limit use of antacids. Routine exercise is beneficial if physician approves.

Pressure ulcers must be dealt with aggressively and promptly. Low hemoglobin has been associated with slow wound healing and hypoalbuminemia.[11] Anemia is also related to hypoalbuminemia. Deficiencies in Vitamin B-12, folic acid, protein, copper, riboflavin, and ascorbic acid exacerbate this problem. The anemia lowers oxygen content of the blood and kills tissues subject to pressure. Chronic infection depresses new red blood cell formation and shortens the life of skin cells. Elevated glucose, dehydration, atherosclerosis, edema, and inadequate food intake also damage tissues. With frequent posture changes, localized treatment, and a diet high in carbohydrates and protein (increase daily protein given from 0.8-1 gram/kg. body weight to 1.2-2.0 grams/kg. unless renal status is compromised [12]), moderate in fat, and supplemented with ascorbic acid and zinc, pressure sores may be deterred or cured. Ascorbic acid and zinc promote collagen formation to maintain good skin integrity. A 35 ml./kg. hydration level is optimal. Small frequent meals become necessary to incorporate these measures. In addition, the following diet patterns may alleviate symptoms for certain patients at the progressive stages or grades of pressure sore or ulcer development:

	Orally Fed Patients	*Tube Fed Patients*
Stage 1	4 oz. Vitamin-C fortified juice at breakfast	Check to see formula order meets RDA for all nutrients
Stage 2	4 oz. Vitamin-C fortified juice at breakfast	Liquid ascorbic acid; increase calories and protein by a factor of 1.1-1.2, check feeding for zinc content
Stage 3	Try 2 eggs at breakfast and 4 oz. orange juice t. i. d. if no calorie restriction imposed	Increase calories and protein by a factor of 1.2-1.5; liquid ascorbic acid
Stage 4	Same as Stage 3 with double portions at one meal if no calorie restriction imposed.	Increase calories and protein by a factor of 2.0

Note that food-based therapy is preferred to supplementation, e.g., ascorbic acid aids in iron utilization; thus, a meal including orange juice and eggs is advised. The use of multivitamin and mineral supplements that are 200 percent of the R.D.A. must be considered in terms of prevention and treatment and may begin at Stage 2. When used, always provide them with meals to maximize their utilization and minimize gastric side effects. Warning: toxic doses appear at ten times the RDA.

Note that a common thread running through the dietary treatments for the geriatric population includes adequate calcium and dietary fiber and moderately reduced levels of total fat, cholesterol, and sodium. It is not too late to intervene in our lifestyles to promote positive nutritional goals such as optimum weight and blood pressure, wound healing, stronger bones and teeth, and reduced risk from cancer and heart disease. Individualized diet counseling, and assessment of individual progress toward nutritional goals with qualified health care professionals, assure us of meeting this achievement.

REFERENCE NOTES

1. Chait, Alan. "Dietary Management of Diabetes Mellitus." *Contemporary Nutrition,* 9(2):1, 1984.

2. Davidson, Mayer B. "Diabetes in the Elderly: Diagnosis and Treatment." *Hospital Practice*, 17(4):114-15, 1982.

3. American Diabetes Association and The American Dietetic Association. "Exchange Lists for Meal Planning." Revised 1989, p. 3.

4. Darby, William J. "Why Salt? How Much?" *Contemporary Nutrition*, 5(6):2, 1980.

5. National Osteoporosis Foundation. "Physicians' Resource Manual on Osteoporosis. A Decision-Making Guide." 2nd ed. National Osteoporosis Foundation, 1991.

6. Henry, H.J., D.A. McCarron, C.D. Morris, and M. Parrott-Garcia. "Increasing Calcium Intake Lowers Blood Pressure: The Literature Reviewed." *Journal of the American Diabetic Association*, 85:182-185, 1985.

7. Heaney, R.P. "Nutritional Factors in Bone Health in Elderly Subjects: Methodological and Contextual Problems." *American Journal of Clinical Nutrition*, 50(5): 1182, November 1989.

8. Shils, Maurice E. and Vernon R. Young. *Modern Nutrition in Health and Disease.* Seventh Edition. Philadelphia: Lea and Febiger, 1988.

9. Public Health Service. *The Surgeon General's Report on Nutrition and Health.* DHHS (PHS) 88-50211. Washington DC: U.S. Government Printing Office, 1988.

10. American Institute for Cancer Research. "Research Update: AICR Study Finds Vegetable Compound Effective Against Cancer." Issue 27, Spring 1990, p. 12.

11. Welch, Trish. "Nutrition-Related Problems in the Institutionalized Elderly." *Dietetic Currents*, 16(1):2, 1989.

12. Litchford, Mary D. "Respecting Nutrient-Ulcer Link Crucial in Treatment Success." *Provider*, 19(3):63, March 1993.

II. Calculation of Oral Intake

Many facilities now require documentation of the percentage of oral intake consumed during meals. After reviewing more complicated methods for a calculation of percentage of foods eaten, I recommend the following 20/20 system:

1. Consider assigning these five food groups 20 points each:

   Meat or meat substitute
   Starch or starch substitute
   Vegetable or soup (use cereal at breakfast)
   Milk
   Dessert or fruit

2. Figure the amount eaten from each food group, multiplying the amount of food left times 20 points. For instance, if one-fourth of the vegetable or soup was not eaten, 5 points of the vegetable group (1/4 × 20 points) was the food left from that group.
3. Add the products from all five groups together to reach the total amount of food not eaten.
4. Subtract this total from 100 points to calculate the total percentage of foods eaten.

Note that total percentage is indicative of total calories only. The points should be tallied from each food group for assessment of more specific nutrient intake. Otherwise, total caloric intake is the only valid parameter that can be evaluated from the total percentage of foods eaten based on the calories contained in a given meal as stated on the menu (see Appendix 10).

III. Nutritional Evaluation of the Elderly

Objective and subjective factors influence the nutritional health of the elderly. As health care practitioners, we must develop accurate, comprehensive assessments of nutritional status. Exploration of the relationship between direct and indirect observations and their standards of measurement help us to generate useful nutritional assessments for the geriatric population.

If the patient demonstrates physical signs of nutrient deficiencies, this finding will be reported in the diagnosis. Physician orders for diet, medications, laboratory tests, and other direct and indirect observations provide a basis for assessment of the elderly population. Comparison of initial and subsequent values for any given modality yields a true evaluation of progress/decline over time.

The initial diet order and subsequent changes in this order should be well-documented in the medical record. Diet orders reflect the type of food items necessary to achieve optimum nutrient levels. If the patient (or significant other) indicates food tolerances/intolerances, these considerations must be communicated to the kitchen staff. The kitchen manager should visit the patient as soon after admission as possible to interview the patient and record preferences in the dietary cardex. Meal rounds assist in gaining insight into patients' specific food habits and appetite levels. A Calorie Count may be ordered (see Appendix 11). This form provides doctors and health care professionals with precise data for assessment in terms of calories and energy nutrients. The form is kept in the patient's room and includes date, meal or snack, food item, and quantity consumed. Results can be calculated using the diabetic exchange system or some other reputable resource material.

CALORIC INTAKE EVALUATION METHODS

Basal energy expenditure (BEE), or the calories the body requires to maintain life at rest, can be derived from the Harris-Benedict equations:

For Men: $66 + (6.2 \times \text{Wt. in Lbs.}) + (12.7 \times \text{Ht. in Ins.}) - (6.8 \times \text{Age in Years})$

For Women: $655 + (4.4 \times \text{Wt. in Lbs.}) + (4.3 \times \text{Ht. in Ins.}) - (4.7 \times \text{Age in Years})$

This equation appears in the Nutritional Assessment form in Appendix 12.[1] Note that the formula uses pounds and inches instead of kilograms and centimeters. Another technique allows 1 kcal./kg. body weight for men and 0.9 kcal./kg. body weight for women (there are 2.2 lbs./kg.). Energy levels of 20%, 30%, 40%, and 50% can be assigned for *very sedentary, sedentary, moderately active,* and *extremely active* activity levels, respectively.

If weight loss is desirable with either method, subtract 500 calories for a 1 lb./week loss or 1000 calories for a 2 lb./week loss (there are 3500 calories in one pound). The opposite applies if weight gain is desirable. Note that caloric needs decline with advancing age (about 10 percent every decade after age 50) because physical activity, lean body mass, and resting metabolic rates decrease with the aging process. Another factor that affects caloric intake needs is fever: for every degree Fahrenheit increase, the metabolic rate increases by 7 percent.[2]

To calculate the grams of each energy nutrient needed for appropriate caloric distribution, multiply the desirable caloric intake by the recommended percentages of carbohydrate, protein, and fat (see previous discussion in *Clinical Topics* section, Chapter I) and divide the subsequent carbohydrate and protein calories by 4 kcal./gram and the fat calories by 9 kcal./gram. The dietetic technician or registered dietitian can then use this data to plan individualized meal patterns. The input from the kitchen manager is most valuable in determining the client's food preferences, which can greatly influence the likelihood of his or her compliance to any given pattern. Most facilities follow a standard meal pattern calculation until data is collected and compiled into a "customized" plan.

BODY SIZE CONSIDERATIONS AND MEASUREMENT METHODS

Height, weight, and other anthropometric (scientific measures of the human body and its parts) characteristics are determined by somatic (pertaining to the wall of the body) protein levels. Therefore, accurate measurements of body size are necessary to ensure that a proper diet is being provided. If a standing height cannot be obtained, there are alternative methods. Stature may be estimated from knee height for both the nonambulatory and those who have spinal curvature. While the elderly individual is lying on his or her back, bend the left knee at a 90° angle and measure the distance from the bottom of the left heel to the patella (above the condyles of the femur). Record the knee height in centimeters (inches multiplied by 2.54) and substitute in this equation:[3]

Estimated Stature (Men) $= 64.19 - (0.04 \times \text{Age}) + (2.02 \times \text{Knee Height})$

Estimated Stature (Women) $= 84.88 - (0.24 \times \text{Age}) + (1.83 \times \text{Knee Height})$

Another estimation method entails measuring the length of the nonambulatory patient's bed and subtracting the distance between the top of the patient's head and the upper edge plus the distance between the heel of the patient's foot and the lower edge.

With these heights one can determine the ideal body weight of some patients using the following convenient methods. Assign 100 pounds for the first five feet of height for a female and 106 pounds for a male. For each inch over 5 feet, 5 and 6 pounds are added for females and males, respectively. If the individual is under 5 feet tall, subtract 2 1/2 pounds for each inch under five feet from the baseline values of 100 and 106 pounds.

These weight results are adjusted downward by 10 percent for a small frame size and increased by 10 percent if the frame size is considered to be large. To determine frame size, place a measuring tape around the smallest part of the wrist distal or towards the fingers to the styloid process of the radius and ulna or "wrist bone." Using the following chart, determine frame size:

Female		*Male*	
Less than 6″	= Small Frame	Less than 6″	= Small Frame
6″ to 6 1/2″	= Medium Frame	6″ to 7″	= Medium Frame
Over 6 1/2″	= Large Frame	Over 7″	= Large Frame

Due to internal body fat gain and lean body mass loss in elderly persons, there is some debate about how to determine ideal weight, but the majority of professionals would agree that this is a reasonable system, especially for the person with diabetes or heart disease. (Regardless, the patient must agree with his or her weight goal to ensure the correct calories are ordered. Despite your best advice to the contrary, the patient may be content at his or her current weight, and documentation should reflect this status.) Furthermore, patients may be classified as obese (20 percent over ideal body weight) when they are not because a person's height declines with age (1/2-1 1/2 inches every 20 years). The third National Health and Nutritional Examination Survey now in progress will provide a more reliable anthropometric reference.

Metropolitan Life Insurance Company increased their ideal weights slightly in 1983 due to lower mortality rates of persons whose weights were slightly higher than the figures given in previously published tables for ideal heights and weights. The company also has an alternative method for measuring frame size.[4] With the arm extended and the forearm bent at a 90° angle, keep fingers straight and turn the inside of the wrist toward the body. Using a caliper (or thumb and index finger), measure the space between the two prominent bones on either side of the elbow. Compare with the following chart. Lower measures indicate a small frame while higher measures denote a large frame.

Height in 1″ heels	*Elbow Breadth*
Men	
5′2″ - 5′3″	2 1/2″ - 2 7/8″
5′4″ - 5′7″	2 5/8″ - 2 7/8″
5′8″ - 5′11″	2 3/4″ - 3″
6′0″ - 6′3″	2 3/4″ - 3 1/8″
6′4″	2 7/8″ - 3 1/4″

Women

4′10″ - 4′11″	2 1/4″ - 2 1/2″
5′0″ - 5′3″	2 1/4″ - 2 1/2″
5′4″ - 5′7″	2 3/8″ - 2 5/8″
5′8″ - 5′11″	2 3/8″ - 2 5/8″
6′0″	2 1/2″ - 2 3/4″

Accurate height and weight readings may also be deterred by edema or water weight, contractures, and amputations. A segmental weight loss of about 10 percent for below-the-knee amputations and 16 percent for above-the-knee amputations may be used with some degree of certainty. An entire arm comprises 6.5 percent and an entire leg 18.5 percent of total body weight.[5] An amputation at the elbow comprises 3.6 percent. Review weights monthly to detect the following significant losses: 5 percent in one month, 7 1/2 percent in three months, and 10 percent in six months.[1] Anything above these amounts is a severe loss.

LABORATORY TESTING IN NUTRITIONAL EVALUATION

Protein-calorie malnutrition can be detected through several laboratory test procedures. Visceral (meaning cavities of the body) protein levels are related to serum albumin and serum transferrin levels, total iron binding capacity (TIBC), and total lymphocyte count (TLC).

Serum albumin levels below 3.3 grams/100 ml. are a sensitive index for protein-calorie malnutrition. Albumin is a protein made by the liver cells. Above-normal values occur due to dehydration, multiple myeloma, and transfusions. Poor protein intake, impaired digestion (including pancreatic insufficiency), and chronic loss of protein (nephrotic syndrome, exuding overhydration) cause some hypoalbuminemia. In the elderly, albumin synthesis is less responsive to an increase in dietary protein; therefore, improvements may be difficult. Two classes of protein, PAB (thyroxine-binding prealbumin) and RBP (retinol-binding protein) assess nutrition status more sensitively due to their rapid turnover (except in liver disease). It is hoped that their costs will be reduced to make their use more practical and pervasive.

Transferrin is an insensitive measure of protein status due to its responsiveness to iron stores, generally elevated in most elderly. Transferrin "transports" iron in the bloodstream on a protein (beta-globulin). Levels below 180 mg./100 ml. indicate pernicious anemia, chronic infection, liver disease, or iron overload. Total iron binding capacity (TIBC) reflects the total amount of iron that can be carried in the plasma by transferrin, and decreased amounts of TIBC correspond to the consequences of lower transferrin.[6]

Total lymphocyte count (TLC) is derived from the number of white blood cells and lymphocytes (infection fighters) by using this equation:

$$\text{TLC} = \%\ \text{Lymphocytes}/100 \times \text{White Blood Cells}$$

If TLC falls below 1500 per mm.,[3] a diminished immune response and subsequent malnutrition might be suspected.

Considerations in Test Analyses for the Elderly

Normal measures of lab values must be carefully interpreted in the case of the elderly. Reduced secretions of digestive juices and enzymes cause some of the nutrients in food passing through the digestive system to remain unmetabolized, thus, unavailable for the body's use. Impairment of bile secretion is quite common; it interferes with normal digestion and utilization of fats, and may result in gaseousness and discomfort after eating fatty foods. Over age 40, normal cholesterol levels may increase to 240 mg./dl. as determined by the National Institute of Health. Red cell count, hemoglobin, and hematocrit standards decline with age and normals range from $4.68 \pm 1.01 \times 10^6$/cu. mm., 14.09 ± 3.43 gm./100 ml., to $42.18 \pm 9.50\%$, respectively.[7] Dehydration tends to elevate hemoglobin as overhydration decreases hematocrit levels. The risk of chronic iron deficiency is increased by the diminished secretion of gastric hydrochloric acid, which leads to a decreased absorption of dietary iron. Osteoporosis has been linked to inadequate calcium intake over a long period. However, blood chemistry values of calcium, phosphorus, and alkaline phosphatase show no specific changes as a result of osteoporosis.[8]

ADDITIONAL CONSIDERATIONS FOR NUTRITIONAL EVALUATION AND CARE

As the elderly are subject to blunted thirst and loss of kidney function, the profound effects of dehydration on fluid and electrolyte balance merit attention. Total body fluid is 60 percent in the elderly as compared with 70 percent in a mature adult. Xerostomia (mouth dryness) interferes with swallowing and contributes to poor intake. Sometimes the elderly restrict fluids to avoid incontinence or frequent urination. Urinary incontinence affects up to 50 percent of the institutionalized elderly. There are indications that extracellular fluid (ECF) declines with age, and climatic heat, fever, infection, and diarrhea also contribute to this problem. Symptoms of dehydration include nausea, constipation, elevated body temperature, hypotension, and mental confusion. Fluid balance must be monitored. Since fluids replace normal losses, aid digestion and intestinal function, and provide for renal clearance, one ml./kcal. or 30 ml./kg. body weight should be available daily.

In completing a skillful nutritional care plan based on assessment, a number of factors should be simultaneously considered. These include physical handicaps (visual, mobility, dental, auditory, dexterity), fluid intake, swallowing capability, skin condition, adaptive eating aids, elimination regularity, appetite level, dining location, food allergies, height compared with weight, food preferences/allergies, estimate of caloric needs, lab data pertinent to nutritional status, drug/nutrient interactions, weight changes, and mental status. Facility requirements vary as to where dietetic professionals document their findings in the medical records, and as to which allied health team meetings would be beneficial for them to attend. Documentation of patient information such as diet order, distinct food likes and dislikes, and snacking patterns should always be kept in the dietary cardex and on the tray card, and updated by the kitchen manager accordingly.

REFERENCE NOTES

1. Blackburn, G.L., Bruce R. Bistrian, Baltej S. Maini, Harah T. Schlamm, and Michael F. Smith. "Normal and Metabolic Assessment of the Hospitalized Patient." *The Journal of Parenteral and Enteral Nutrition*, 1(1):13-14, 1977.

2. University of Washington. *Manual of Medical Therapeutics.* 27th Edition. Boston: Little, Brown and Company, 1992.

3. Roche, A.F., R.M. Siervogel, and W.C. Chumlea. *Nutritional Assessment of the Elderly Through Anthropometry.* Yellow Springs, OH: Fels Research Institute, 1984, pp. 6, 10.

4. Metropolitan Life Insurance Company (derived primarily from data of the 1979 Build Study). Society Of Actuaries and Association of Life Insurance Medical Directors of America, 1980.

5. Brunnstrom, S. *Clinical Kinesiology*, Philadelphia: F.A. Davis, 1979.

6. Fischbach, Frances. *A Manual of Laboratory Diagnostic Tests.* Third Edition. Philadelphia: J.B. Lippincott, 1988.

7. Freedman, Michael L. "Anemias in the Elderly: Physiologic or Pathologic?" *Hospital Practice*, 17(5):122, May 1982.

8. Spencer, Herta. "Osteoporosis: Goals of Therapy." *Hospital Practice*, 17(3):138, March 1982.

IV. Adaptive Eating Devices

Elderly clients will benefit at mealtimes if they are able to handle food easily. The objective is to facilitate maximum self-feeding independence. If trouble exists with utensil handling because of manual disability, a simple aid may help. A few of these aids are discussed here by their functional purposes.

For those with severely limited motion in the wrist and fingers, scoop-shaped plates help catch food on eating utensils. Snap-on food guards can be obtained that fit onto regular 9- to 11-inch plates. Silverware comes with right- or left-handed spoons with angled shanks, and a knife that is used with a rocking-slicing action. This rocker knife enables food cutting with one hand. There is even a mug cut out at the bottom so one's hand slides in to pick it up.

Eating utensils with built-up grips aid the arthritic or stroke-stricken individuals who cannot grip items easily. To achieve the same effect, a foam rubber curler can be placed over regular silverware, if preferred. If grasp is almost completely lacking, a palmar cuff, consisting of a band with a Velcro fastener, adjusts to fit the hand. Elastic webbing is also available. Silverware the fits easily into the palmar pocket worn over the knuckles. A loss of grasp can also be overcome using a mug with a T-handle that is easy to pick up by a variety of grasping patterns.

In cases of marked weakness or poor coordination of the upper extremities, it is usually easier to sip through an ordinary straw rather than trying to pick up a glass. When poor coordination or difficulty rotating the forearm become bothersome, the use of a swivel spoon is helpful; it rotates to keep the bowl of the spoon level as it is lifted to the mouth. Forks, soup spoons, and even sporks are also available.

Messiness seems to become an especial problem with drinking. For those with a shaky or wobbly grip, try cups with limited cut-out areas that allow tipping without bending the head back. Cups with spouts or bent mouthpieces (similar to built-in straws) may prove useful. There is a one-way drinking straw that remains full even after suction stops. Dysphagic, severe arthritic, and Parkinson's disease patients may benefit from it.

If dishes tend to slide for any reason while eating, try placemats made of a nonslip material or suction cup pads. Some have even found a damp paper towel or sponge cloth under the plate works wonders.

Divided plates assist the visually impaired in locating foods by using the "clock method." The divisions of the plate can be traced to note food arrangement, e.g., peas at 10 o'clock, carrots at 2 o'clock, meat at 6 o'clock.

In addition to using adaptive eating devices, some facilities have established therapeutic or restorative dining rooms to provide adequate individualized attention for clients requiring assistance.

A consultation with an occupational therapist may yield great assistance in determining solutions to clients' needs.

V. Drug-Nutrient Interactions

The scope of food-medication reactions plays a role in food intake, nutrient absorption, nutrient metabolism, nutrient excretion, absorption of drugs, and drug action. Specific examples of how certain categories of drugs affect nutrients, or if the reverse situation is applicable, how the nutrients affect drug reactions (usually adversely), allow better insight into the requirements for a comprehensive geriatric nutrition care plan.

The implications of drug-nutrient interactions become meaningful for two reasons: (1) the long-term uses of medications by persons as they age, and (2) the dramatic toxicity and efficacy of pharmacokinetics[1] (pharmokinetic means that one drug interferes with the absorption or disposition of another). If the geriatric individual is being treated for more than four pathological entities, it is likely that multiple drug regimens may be ordered by the physician. (About 45 percent of elderly patients take multiple prescriptions.) It is possible that a condition such as malnutrition might remain undetected if drug side effects mimic its symptoms. These therapeutic effects result from a number of causes.

Diminishment of lean body mass in relation to total body weight results in higher drug-to-weight ratios when standard doses are used. Decreased stomach acid and slow transit time make blood levels after standard doses as high in elderly patients as in young patients. Decline in cardiac function produces delays in drug arrival at the site of action and delays in removal from the site.

Most drugs bind to serum protein, but a decline in this protein in the aged elevates blood levels of free drugs. Decreased numbers of nephrons in the kidney, decreased renal blood flow, and decreased creatinine clearance delay drug excretion even when normal BUN (blood urea nitrogen) and creatinine levels exist. The liver's ability to metabolize and detoxify enzymes declines, yielding potentially elevated serum levels of active drugs.[2] Drug distribution is also affected by a higher ratio of fat to protein, which increases the storage room for fat soluble drugs.[3]

When drugs accumulate in the bloodstream, altered brain function is often the earliest warning signal. Dementia has even been mistaken for B-6, B-12, and folacin deficiency. As activity levels increase, the body absorbs and eliminates drugs faster. Thus, metabolic changes influence this situation also.[4]

Prescribed drugs most commonly fall into four general categories. The following chart lists these categories and others used to treat geriatric problems. Included are: over-the-counter and by-prescription drugs with their appropriate uses; and nutritional interactions and suggested dietary therapies.

CATEGORY	USE	NUTRITIONAL INTERACTION	THERAPY
ANTI-ARTHRITIC:			
Aspirin	Anti-Inflammatory	Enhanced GI Bleeding	Do not give in Vitamin K deficiency and increase iron-rich foods
Ibuprofen		Promotes iron-deficiency anemia and possible kidney damage	Provide 6-8 glasses water daily
Cortico-steroids		Interferes with bone metabolism and nutrient absorption although increases excretion	Consider calcium (or even vitamin/mineral) supplements, then provide 6-8 glasses of water daily to prevent kidney stones. Increase protein to offset catabolic properties over the long term

CATEGORY	USE	NUTRITIONAL INTERACTION	THERAPY
GASTRO-INTESTINAL:			
Aluminum-containing antacids	Hyperacidity	Lower gastric acidity; constipation; calcium loss	May need elevated calcium intake
Magnesium hydroxide		Laxative effect; elevated magnesium levels in patients with renal failure	Same as above
Calcium carbonate (contains 40% calcium by weight)		Constipation	Give one to three hours after meals
Sodium Bicarbonate		Altered blood pH; systemic sodium load	Chronic use should be discouraged
Psyllium Preparations	Bulk formers	May lower cholesterol; contains dextrose kcal	Give with large volume of water
Mineral Oil	Lubricant	Decreased absorption of fat-soluble vitamins–carotene and Vitamins D, E, and K–when taken with food. No effect on caloric intake; indigestible	Take two hours away from food on empty stomach
Olive Oil	Laxative	Increased caloric intake; digestible	Same as above

CATEGORY	USE	NUTRITIONAL INTERACTION	THERAPY
Docusate Sodium/Calcium		Altered intestinal absorption of electrolytes	Give with milk or fruit juice
Natural Vegetable Compound		Gastric irritant	Give at H.S. with eight ounces of water
Bisacodyl		Lowers glucose absorption	Give with eight ounces water on empty stomach[5]
CARDIOVASCULAR PREPARATIONS:			
Anti-hypertensives such as diuretics, etc.	Reduce blood pressure	Elevated sour and salt thresholds	Give with meals or milk
Diuretics except spironolactone, triamterene, or amilorid		Hypokalemia; sodium, zinc, and magnesium balance adversely affected	Avoid natural licorice.[6] Provide K+ and Mg++ rich foods. Monitor fluid intake. Avoid strict low sodium diet and isotonic formula feedings.
Glycosides such as digitalis or digoxin	Cardiac insufficiency	Loss of appetite and urinary potassium and calcium	Avoid natural licorice; maintain high K+ diet
Bile Acid Sequesterance such as cholestyramine or colestipol	Reduce serum cholesterol	Decreased absorption of fat soluble vitamins, B-12, and folate	Consider a multivitamin[7]
Nicotinic Acid		GI upset	Give with food

CATEGORY	USE	NUTRITIONAL INTERACTION	THERAPY
PSYCHOACTIVE:			
Sedatives	Anxiety, tension	Increased Vitamin C requirement; deficiency in folate and Vitamin D; effect altered by alcohol	May require vitamin supplements; avoid alcohol
Benzo–diazepines		Significant weight gain in some, however may cause disinterest in food among older pattients	Limit intake of coffee and tea; give with food
Phenothiazines	Antispychotic	Increased appetite	Monitor oral intake; give with food
Thioridazine		Increased appetite, however, over sedation causes loss of appetite	Monitor oral intake; give with food
Amitriptyline	Depression	Interferes with eating by causing agitation of some elderly	Monitor[8]
OTHERS:			
Anti-convulsants	Seizure disorders	Can interfere with B-12, folacin, and calcium (bone metabolism)	Consider calcium supplement, but provide 6-8 glasses of water daily if given

CATEGORY	USE	NUTRITIONAL INTERACTION	THERAPY
Antidiabetic Agents	Diabetes mellitus	Allows for glucose utilizaiton	Avoid alcohol
Antibiotics	Bacterial infections	Can increase stool volume, frequency, and quantity of undigested food	Give on empty stomach; do not give with fruit juices
Tetracycline			Avoid milk products
Parkinsonian Drugs	Parkinson's Disease	Effects decreased by high protein intake; interferes with B-6*	Give low protein diet, possibly as low as 0.5 gm. per kg.
Unpleasant Tasting Drugs		Alter taste sensation; decrease food intake when used chronically	Offer with applesauce or fruit juice as indicated[9]

*Best sources of B-6 are meat, especially liver; vegetables including potatoes; wheat bran; and whole grain cereals.

Adverse nutrient reactions due to medication administration is not inevitable. Selection of a nutritionally well-balanced diet from a wide variety of foods prevents depletion of vitamins and minerals if overall nutritional status is good.

ASPIRIN AND DIURETICS

Significant dietary interactions can be attributed to two of the most widely used medications: aspirin and diuretics.

Aspirin

Aspirin may seem an unlikely target for concern due to its over-the-counter availability as an analgesic, anti-pyretic, anti-inflammatory, and blood clot-preventing agent. The latter feature appears to aid in reducing the risk of heart attacks or recurrent myocardial infarction, and explains the association between aspirin use and fewer strokes related to blood clotting. However, hemorrhagic strokes, which are more likely to be fatal, occur with aspirin *because* of its blood clot-preventing feature.[10] The risk of cancer of the stomach, esophagus, colon, and rectum declines with regular, prolonged use of aspirin.[11] Aspirin is also a known gastric irritant that lowers the stomach's pH (pH is a measure of body acidity or alkalinity). When aspirin is being used, acidic foods should be avoided, such as caffeine, citrus fruits, cola drinks, fruit juices, pickles, tomatoes, and vinegar.[12] In combination with any one of these foods, aspirin can corrode the stomach lining and cause ulcers with prolonged usage.

Aspirin dissolves best in a higher pH environment. Buffered aspirin (ASCRIPTIN®, BUFFERIN®) contains antacids or buffering agents and elevates the pH in the diffusion layer to enhance the diffusion and absorption rates.[13] (However, it may still cause irritation.) Many elderly patients receive aspirin in this form. Effervescent-type aspirin solutions (ALKA SELTZER®) achieve buffering action at the expense of a high sodium content. They are highly absorbable since already in solution. Various types of aspirin contain approximately 50 mg. of sodium (Na) per dose; effervescent types with over 500 mg. Na per tablet[14] are not recommended for diets that must limit salt intake. Aspirin is absorbed most efficiently in the small intestine rather than the stomach, due to the small intestine's higher pH level.

Doctors recommend that aspirin be ingested immediately before, with, or after a meal or with a snack (with the exception of the aforementioned foods) to prevent or minimize stomach upset. Those who experience marked epigastric or abdominal distress should especially take this medication with food. Enteric-coated aspirin preparations that are absorbed in the lower small intestine are probably less well-absorbed than those without the coating.

Enteric-coated (ECOTRIN®) and sustained release preparations (MEASURIN®) may not release medication as expected when mixed with food. Therefore, it is better to take them on an empty stomach. Note that enteric-coated capsules should definitely not be taken within one hour of milk ingestion. In fact, any food containing milk may retard the enteric type's performance, and milk may enhance dissolution of the enteric coating which is designed to protect the stomach from irritation.[15]

Also, aspirin should be taken with a full glass of water (240 ml. or 8 oz.) to promote dispersion and dissolution of the drug, to maximize bioavailability (readiness for the body's use), and to reduce acid degradation of the esophagus and stomach. Generally, aspirin should not be taken within one hour of bedtime, when the possibility of gastric reflux is greatest. If a danger of GI bleeding exists, most aspirin may be taken with milk or dispersed in two tablespoons of antacid to enhance dissolution in the stomach. Although aspirin absorption is delayed with food ingestion, its bioavailability remains unaffected or only slightly reduced by the presence of food.

Diuretics

Historically, diuretics have been utilized for weight loss programs, but, of course, modern medicine frowns on this practice. Loss of fat (not water) indicates true progress for the overweight or obese. Xanthine diuretics, members of the caffeine family, have been supplanted by more potent compounds. Tea and coffee produce a partial diuresis (increased urine volume). So-called diuretic teas such as buchu,[16] quack grass, and common dandelion are available, but their diuretic action is probably no greater than that of the xanthine alkaloids of coffee or ordinary tea. Toxic effects have been reported from juniper berries and shave grass (or horsetail plants), two ingredients in teas with diuretic action. The former can irritate the gastrointestinal tract. Nicotine, contained in the latter, may cause excitement and loss of appetite. Of utmost significance, agressive diuretic therapy in older patients is more likely to cause dehydration and hypovolemia (diminished blood volume) due to older persons' lack of water reserve. Some safer and more effective alternatives are discussed here.

The most common form of diuretics, called thiazide derivatives, increase the excretion of water by the kidneys with a concurrent loss of sodium. There are now more than ten of these agents in use. These drugs and more potent ones possess the ability to deplete the body of water, sodium, potassium, zinc, copper, and magnesium to such an extent that a vitamin/mineral supplement is recommended daily. The "loop" diuretics, named for their inhibition of sodium and water reabsorption at the Henle's loop in the renal tubule(s) are the strongest and fastest-acting. To compensate for potassium loss, high potassium foods should be served. The best sources are: bananas, prunes, oranges, grapefruit, dried apricots, figs, peaches, dates, cantaloupe, watermelon, yams, squash, dry lentils, broccoli, white and sweet potatoes, brussels sprouts, cauliflower, tangerine and pineapple juice, all fresh fish, veal, chicken (white meat only), beef liver, beef, and pork. Excellent magnesium sources include whole grains, nuts, beans, and green leafy vegetables. Furosemide (LASIX®), a loop diuretic, also leads to depletion of calcium from the body.[17]

With the advent of the potassium-sparing diuretics–spironolactone (aldactone®), triamterene (dyrenium®), and amiloride (midamor®)–a more effective way of conserving body potassium has been discovered. These drugs are more easily tolerated and more reliable than potassium supplements for elevating serum potassium during diuretic therapy. They are usually used in conjunction with the thiazide diuretics.[18]

No diuretic is able to reduce salt and water in body tissues effectively if large amounts of salt are being added to the diet. For maximum benefit, a doctor's advice should be followed. Most Americans consume 5-15 grams of salt daily. This can be reduced to below five grams by avoiding table salt, cured and processed meats, condiments (catsup, mustard, soy sauce), salted snacks (potato chips, pretzels, salted nuts, and crackers), sauerkraut, pickles, olives, broth, and canned or dehydrated soups. Herbs and spices used in place of salt enhance the flavor of most foods appreciably. All fruit is low in sodium. Fresh fruits and juices (e.g., lemons, oranges) can be used as food garnishes.

Monosodium glutamate (MSG) and natural licorice should be avoided during diuretic therapy. MSG also removes excess water from body tissue, and in combination with a diuretic may deplete essential quantities of water-soluble vitamins (C and B-complex) and minerals (sodium and potassium). Exercise caution with natural licorice also. Most licorice in this country is artificial and not harmful, but licorice candy and flavoring imported from Europe is often natural. A steroid in natural licorice root causes salt and water retention and hypokalemia that can lead to increased blood pressure and heart attacks.[19]

REFERENCE NOTES

1. NIH. "Our Future Selves: Report of the Panel on Biomedical Research." NIH Publication No. 80-1445, Washington, DC, 1980.

2. NIH. "Observations on Geriatric Medicine." DHEW Publication No. (NIH) 79-162, Washington, DC, 1979.

3. NIH. "Changes. . . Research on Aging and the Aged." NIH Publication No. 81-85, Washington, DC, 1980.

4. Hickey, Tom. *Health and Aging*. Monterey: Brooks/Cole Publishing, 1980, p. 114.

5. Decker, Edward L. "Interactions: The R.D. and the M.D," *RD*, 1(3): 2,3; 1981.

6. Powers, Dorothy E. and Ann O. Moore Allen. *Food-Medication Interactions*. 7th Edition, 1991. Tempe, AZ: Ann Moore Allen.

7. Roe, Daphine A. "Drug and Nutrient Interactions in Elderly Cardiac Patients." *Drug-Nutrient Interactions*, 5: 205-12, 1988.

8. Medical Economics Data, editor. *Physician's Desk Reference*. 47th Edition, Montvale, NJ: Medical Economics, 1992.

9. Grant, Anne and Susan DeHoog. *Nutritional Assessment and Support*. 4th Edition. Seattle: Anne Grant/Susan DeHoog, 1992.

10. Herbert, Victor. "Iron Deficiency in the Elderly." *Nutrition & the M.D.*, 16(10):1,2, October 1990.

11. Thun, Michael J. et al. "Aspirin Use and Risk of Fatal Cancer," *Cancer Research*, 53(6):1322-1327, March 15, 1993.

12. Benowicz, Robert J. "Foods and Drugs You Mustn't Mix." *Woman's Day*, July 17, 1979, p. 66.

13. American Pharmaceutical Association. *Handbook of Nonprescription Drugs*. Seventh Edition. Washington, DC: American Pharmaceutical Association, 1982.

14. United States Department of Agriculture. "The Sodium Content of Your Food," *Home and Garden Bulletin*, No. 23, Washington, DC, 1980.

15. Tuttle, C. Brian. "Harmony with Drugs and Food." *Canadian Medical Association Journal* 126(10): 1161-1162, May 15, 1982.

16. Brunner, Lillian S. *Textbook of Medical-Surgical Nursing*. Sixth Edition, Philadelphia: J.B. Lippincott, 1988, pp. 579-80.

17. Spencer, H., L. Kramer, and D. Osis. "Factors Contributing to Calcium Loss in Aging." *American Journal of Clinical Nutrition*, 36:776-787, 1982.

18. Rosamond, Thomas, L. and Larry E. Fields, "Hypertension," in William Claiborne Dunagon and Michael L. Ridner (eds.), *Manual Of Medical Therapeutics*. 26th Edition. Boston: Little, Brown, and Co., 1989, pp. 78-80.

19. Powers, Dorothy E. and Ann O. Moore Allen. *Food-Medication Interactions*. 7th Ed. Tempe, AZ: Ann Moore Allen, 1991, pp. 6, 189.

VI. Modified Food Texture for Swallowing Disorders

Approximately 40 percent of long-term care patients require pureed or mechanically-altered foods. The most obvious cases include those with loss of teeth, ill-fitting dentures, diminished gag reflex, extreme mouth soreness, and esophageal stricture. Some clients just simply do not have the mental capacity to remember to chew and swallow. Any chewing or swallowing disorder, regardless of the cause, is referred to as dysphagia. It may result from cerebrovascular accident (CVA); from degenerative neurological disease such as dementias, Parkinsonism, amyotrophic lateral sclerosis, multiple sclerosis, and myasthenia gravis; or from laryngeal and oral cancer.

Some dry foods may be offered to patients who have enough tongue movement and saliva to turn and to moisten them; e.g., plain, fresh cake with icing usually requires no blending. Those with poor tongue control benefit from pureed foods. Various degrees of liquified substances are utilized to meet different needs. Thin liquids often prove less tolerable than thicker, blenderized items.

Swallowing refers to the entire act of deglutition, from placement of food in the mouth, through the oral and pharyngeal stages, until the material enters the esophagus. No age-related change in the oral phase of the swallow has been noted; however, there is some reduction in the strength of the pharyngeal peristalsis, resulting in residual material in the pharynx after the swallow (but this generally does not occur until the eighties).

The following stages outline a progression of food consistency levels to be used if a delayed or absent swallowing reflex exists.[1]

Stage 1: NPO–Single items ordered with physician's approval for trial basis. (A speech/swallow evaluation is strongly recommended.)

Stage 2: PUREE–No liquids orally unless approved through physician's order. Smooth, moist, and blenderized foods only.

Stage 3: "MECHANICAL SOFT"–A much more specific term than just "chopped," this designates no liquids unless specified by physician's order. Finely ground, casserole, and soft foods that require little chewing effort. Hold ground meat on serving line in gravy for improved tenderness. If ratio of 1 oz. gravy: 2 oz. meat formulated, a 6 oz. ladle to serve 3 oz. meat must be utilized. Gravies and sauces necessary. No breads, acidic foods, peas, dry cereals, baked potatoes, peanut butter, fried foods, or hard cheeses.

Stage 4: THICK LIQUIDS–Thickened, dense liquids that cannot pass through a straw, such as farina, thickened cream soup, and smooth yogurt.

Stage 5: MEDIUM LIQUIDS–Includes eggnogs, thin milk shakes, regular cream soups, and nectars.

Stage 6: THIN LIQUIDS–Progression to clear liquids, Italian ice, and milk. No straws because they inhibit swallowing control.

It is instructional to note that 47 percent of those over age 65 take medication that can diminish salivary flow. Saliva lubricates oral soft tissues and food, provides debridement of tooth surfaces, influences plaque accumulation, and distributes bacterial acids producing caries.[2] Antidepressants and tranquilizers may cause xerostomia (dry mouth) in addition to sedation and lack of coordination.

In facilities fortunate enough to have a restorative feeding program comprised of a team of experts (who provide synergy to each other)–physician/radiologist, speech therapist, nurse, and dietitian–the initiation and progression of food consistency levels will be greatly facilitated.

The following are some guidelines for temporary diet management in response to complications in the dysphagia patient[3]:

- Regurgitation of food through nose (nasal regurgitation): physician must be advised immediately.

- Reduced attention and orientation; distractibility: prompt verbally.
- Behaviors such as impulsivity affecting eating rate and amount consumed per mouthful: ample supervision necessary.
- Abnormal head and body position: nursing support mandatory.
- Slow eating: provide ample time; consider six small meals.
- Chewing difficulties: decrease the texture stage of the problem item(s).
- Drooling of liquids or bolus (partially chewed/digested foods): control with spoon-thick liquids (refer back to Stage 4) and blenderized textures.
- Pocketing of food in the sides of mouth, on the roof of mouth, and/or under the tongue: decrease texture by one stage (a finger sweep after meals clears any pocketed food).
- Coughing or choking: popsicles, frozen slushes or blenderized frozen juice, spoon-thick liquids (Stage 4): decrease texture by one stage.
- Changes in vocal quality following swallowing (examples: hoarseness, wet, or gurgly quality): *evaluate* for tube feeding if difficulty in breathing, continued coughing, and need for suctioning present.

Patients with weak or uncoordinated swallows, decreased laryngeal elevation, poor oral muscular control, or reduced oral sensations may benefit from semisolids instead of purees to trigger the weak reflex and maximize the stimulus for swallowing.[4] Of course, a feeding tube for the neurologically impaired is recommended, since they are at high risk for aspirating depending on the progression of the disease. (Aspiration is entry of foreign material into the airway below the true vocal cords. This is dangerous because fluid that forms around the foreign substance is conducive to pathogens, meaning, for example, development of pneumonia.)

The challenge to the dietary department is to prevent the complications of dysphagia–pressure sores, protein-calorie malnutrition, nasal regurgitation, aspiration, and depression. Present the pureed entrees in serving bowls designed for the amount given to avoid runny appearance and thermal loss. Thickening prevents foods and beverages from free-falling through the throat onto the vocal cords or into the lungs. Foods can be shaped to resemble their natural counterparts and garnished with gelatin cubes, aspics, Parmesan cheese, and dollops of sauces such as smooth cranberry, mayonnaise, whipped cream or cream cheese, sherbet, apple butter, jellies, and spices. Good menu planning as discussed earlier is even more important in dealing with swallowing disorders. To add calories without bulk (i.e., to make calorie-dense preparations), add margarine to hot dishes; cream, evaporated, or half-and-half milk instead of water to soups, puddings, or cocoa; sour cream in twice-baked potatoes; and powdered coffee cream or dry milk to milk shakes, hot cocoa, hot cereal, gravies, and soups. Remember, the thicker the dish, the more calories per serving; runny or watery items contain far fewer calories per serving.

Possibly the toughest challenge in caring for patients with dysphagia is providing enough dietary fiber to meet the recommended 20-30 grams daily. The addition of processed bran (see suggestions in the "Specialized Recipes" chapter) will help to meet the necessary high fiber and calorie-dense standards.

REFERENCE NOTES

1. Cherney, Leora, Jean Pannell, and Carol Cantieri. *Clinical Evaluation of Dysphagia.* Gaithersburg, MD: Aspen Publications, 1985.

2. Carlos, James P. and Mary D. Wolfe. "Methodological and Nutritional Issues in Assessing the Oral Health of Aged Subjects." *American Journal of Clinical Nutrition*, 50:1210, 1989.

3. Kaskel, Phyllis. "Dysphagia." *The Journal of Applied Nutrition*, 1(10), August 1987.

4. Curran, Jean and Michael Groher. "Development and Dissemination of an Aspiration Risk Reduction Diet." *Dysphagia*, 5:7, 1990.

For further reading, strongly recommended is: *Dysphagia: Diagnosis and Management*, 2nd Ed., by Jean Curran, 1991. Available from Butterworth (Boston).

VII. Fiber for Thought

Plant fiber abounds in fruits, vegetables, legumes, and grains such as rice, wheat, rye, barley, oats, and corn. The most efficient sources are the cereal brans or outer husks of these grains. Scientists call the plant constituents which are resistant to digestion, as measured by the secretions in the human gastrointestinal tract, dietary fiber. Researchers have "refined" their knowledge of fiber enough to differentiate between soluble and insoluble types. The benefits of both varieties and ways to include them in daily diets are discussed.

Soluble fiber has high water-retention capacity and turns to gel during digestion. The consequent slowed absorption and transit time from the stomach to intestine affects cholesterol absorption. Pectins from fruits and vegetables, gums found in oat and rice bran, plus lentils, barley, cornmeal and seeds are examples of soluble fiber. Don't forget either chili (kidney beans) and cornbread or baked beans (great northern). Although soluble fiber accounts for just 25 to 35 percent of fiber intake, it has achieved national acclaim for its role in cholesterol control.[1] Soluble fiber supplements have been shown to lower serum low-density lipoproteins (LDLs), cholesterol levels.[2,3] Coronary heart disease is characterized by deposits of these lipids (composed mainly of cholesterol) that accumulate in the large and medium-sized arteries serving the heart. Soluble fibers have also been associated with improved sugar control in diabetics.

Interestingly, monosaturated oils–olive, peanut, canola, and avocado oils–actually lower cholesterol levels when used in moderation. Polyunsaturated fats such as safflower, sunflower, corn, and cottonseed oils were formerly believed to be the only ones that lowered serum cholesterol.[3]

The cellulose and lignin of insoluble fiber speed the passage of foods through the stomach to the intestines, providing bulk to the diet. It has poor water-retention capacity. Wheat bran intake has been linked with a decreased incidence of colon tumors and cancer.[4]

The National Cancer Institute recommends 25-35 grams of total dietary fiber daily, yet, most Americans consume scarcely half this much. The American Cancer Society recommends foods from the cruciferous family, such as cabbage, broccoli, brussels sprouts, kohlrabi, and cauliflower, to help prevent cancer. Include plenty of vegetables and fruits: choose Vitamin A sources such as carrots, spinach, sweet potatoes, cantaloupe, and apricots, and Vitamin C sources such as citrus fruits (oranges, grapefruits), strawberries, and green and red peppers. These foods may deter cancers of the larynx, esophagus, and lung. A correlation of diets high in fat and calories to breast and colorectal cancers prompts the recommendation that skim milk be used whenever possible.

NOTES

1. Feldman, Elaine B. "Know Your Cholesterol–and the Proper Diet." *Nutrition & the M.D.*, April 1988, p. 2.

2. Marlett, Judith A. "NCEP Announces New Guidelines for Lowering Blood Cholesterol and Coronary Risk." *Oat Fiber Factor*, 1:4, 1988.

3. Kowalski, Robert E. *The Eight-Week Cholesterol Cure*. New York: Harper and Row Publishers, 1987, pp. 51, 87.

4. Monmaney, Terence et al. "The Cholesterol Connection." *Newsweek*, 111 (6): 56, February 8, 1988.

5. National Cancer Institute. "Diet, Nutrition & Cancer Prevention: The Good News." NIH Publication No. 87-2878, December, 1986.

VIII. Calcium and Osteoporosis

Statistics show that 90 percent of elderly Americans who sustain a broken hip never recover.[1] The importance of this stems from the fact that even the slightest amount of stress causes fractures in the elderly due to weak bone structure. The hip, spine, and wrist deteriorate the most. When the alveolar ridge under the gums declines, dentures do not fit properly. Dowager's hump, a gradual loss of height with spinal curvature, occurs due to bone loss. Bone fracture under everyday stress and bone deterioration are related to a condition termed osteoporosis, or porous bones.[2,3]

Physicians may prescribe calcium supplements of at least 1500 mg. daily or estrogen therapy to replenish calcium stores.[4] Dietary habits serve as a useful tool for providing an adequate calcium supply. About 99 percent of the body's calcium is stored in the bones and teeth while the other 1 percent stays in the bloodstream. This small amount regulates muscle contraction and relaxation, blood coagulation, transmission of nerve impulses, activation of enzymes, stimulation of hormones, and integrity of intercellular cement substances.[5]

Although the Recommended Dietary Allowances now allow for greater calcium intake (1200 mg.) among 11- to 18-year-old females and pregnant and lactating women, the National Institute of Arthritis, Diabetes, and Digestive and Kidney Diseases recommends 1000-1500 mg. of calcium daily for post-menopausal women.[6] Dietary calcium and its helpers in bone growth–Vitamin D and phosphorus–are easily obtained from four 8 oz. glasses of Vitamin D-fortified milk daily. For those who do not like milk, substitutions for one cup of milk include: 1 cup yogurt, 1 1/2 ounces of cheddar cheese, 1 cup pudding, 1 3/4 cups ice cream, or 2 cups cottage cheese. For those with lactose (milk sugar) intolerance or milk sensitivity, Swiss and cheddar cheese contain the lowest lactose content. Consumption of milk with other foods is less likely to produce problems, and lactose-reduced milk contains about 70 percent less lactose than regular milk. An alternative is to add the enzyme, lactase, to regular milk. Lactase enzyme products digest the sugar in milk, and are available over the counter at any pharmacy.

Research has shown that 1000 mg. of calcium daily lowers blood pressure in normal and hypertensive human subjects.[7] An above-average intake of calcium may also result in a lengthened interval between the attainment of maturity and the onset of senility. Some studies even indicate that an increased intake of calcium decreases the likelihood of colon cancer. Apparently, two-thirds or more of ingested calcium remains unabsorbed and binds with fatty acids from food in the lining of the large bowel; this prevents the acids from irritating the colon.[8] Also note that too much caffeine in the form of soft drinks, coffee, and tea causes bones to lose calcium. (There is now evidence that the phosphorus in soft drinks can disrupt the calcium-phosphorus ratio of 1:1 in the bloodstream if ingested in large amounts.)

Popular recipe favorites such as broccoli-cheese casserole, double cheese pizza, cheese soup, and custards should become a vital part of daily menus. Frozen yogurt and cheddar cheese wafers make delicious alternatives. Serve cheese grated as a garnish for chilies and tossed salads or in cubes as an accompaniment to fruit. Low fat (reduced calorie) brands are wonderful choices.

It is never too early or too late to begin selection of a better diet. Let's "break" old habits, not old bones!

REFERENCE NOTES

1. National Institute of Arthritis, Diabetes, and Digestive and Kidney Diseases. "Osteoporosis: Cause, Treatment, Prevention." NIH Pub. No. 83-2226, April 1983.

2. Heaney, R.P., J.C. Gallagher, C.C. Johnston, R. Neer, A.M. Parfitt, and G.D. Whedon. "Calcium Nutrition and Bone Health in the Elderly." *American Journal of Clinical Nutrition*, 36:986-1013 (Supplement), 1982.

3. Allen, L.H. "Calcium Absorption and Requirements During the Life Span." *Nutrition News*, 47:1, February 1984.

4. Avioli, Louis. "Osteoporosis: Diagnosis and Treatment." *Myology*, 4:2, 1979.

5. Shils, Maurice E. and Venon R. Young. *Modern Nutrition in Health and Disease.* Philadelphia: Lea and Febiger, 1988.

6. Food and Nutrition Board. *Recommended Dietary Allowances.* Tenth Revised Edition. Washington, DC: National Academy of Sciences, National Research Council, 1989.

7. Morris, C.D., H.J. Henry, and D.A. McCaron. "Randomized, Placebo Controlled Trial of Oral Ca 2+ in Human Hypertension (Abstract)." *Kidney International*, 25, January 1984.

8. Barger-Lux, Janet M. "The Role of Calcium in Preventing Diseases of Aging." *Provider*, 16(2):33-4, February 1990.

IX. Enteral Feeding

The tube- or enteral-feeding client usually needs 21-43 ml. fluids/kg. body weight. Formulas are mostly water (about 85 percent) and can be included in the total fluids given daily. The National Academy of Sciences-National Research Council (NAS-NRC) suggests one ml. water/kcal. for adults and 1.5 ml. water/kcal. given to infants under normal circumstances. Fluid requirements become increased through voluntary or involuntary water loss: as a result of fever, or based on neuromuscular control. Special problems such as dumping syndrome, diarrhea, bloating, and retention can be addressed through reduced rate of tube feeding flow. The lowering of osmolality may help alleviate the first and second problems. Warming the feeding to room temperature is advisable. Continuous feeding is widely used to avoid introduction of air into the gastrointestinal tract, and to prevent diarrhea caused by bolus feeding, delayed gastric emptying, and recurrent pulmonary aspiration.[1]

Specialized formulas meet specific needs. A pharmaceutical representative or a diet manual may be very helpful in determining the necessary protocol. Some of the liquid supplements assist regular bowel functioning by providing three grams of dietary fiber per 240 ml. or cc. Avoid antibiotics, sorbitol in elixir-type medications, or antacids when diarrhea occurs.[2] Minimize this situation by offering lactose-free isomolar formulas. Duodenal feeding tubes are still preferred for long-term usage in patients. Usually, a calorie-nitrogen ratio of 150:1 suffices. A 100:1 ratio product or a modular protein powder may be utilized for those in need of protein repletion or weight gain.[3] Nitrogen balance, calories ingested, and anthropometric measures must be monitored on a regular basis (see Appendix 13).

It is important to provide accurate feedings regardless of the strength or type of formula ordered. The following methods give specific examples for calculations:

METHOD FOR CALCULATION OF TUBE FEEDING STRENGTH USING A LIQUID FORMULA

Step 1. From the physician's order, determine the fractional strength, e.g., 1/4, 1/2, or 3/4.

Step 2. If the physician's order for the formula is in fluid volume, skip to Step 3. If the physician's order is in total kcal., figure the fluid volume of those kcal. by dividing the kcal./ml. or cc. for full strength into the total kcal. ordered.

Step 3. Divide the numerator of the fraction of strength into the formula volume ordered to get a *factor*.

Step 4. Multiply this factor times the denominator. This number represents total volume including both formula and water.

Step 5. Subtract the formula volume of Step 2 from the total volume of Step 4. This number is the amount of water that must be added to the fluid volume of Step 2 to achieve the strength required.

Note: If the order reads ml./hr., a simpler method applies. Just multiply by the correct fractional strength to calculate the volume of formula/hour. Subtract this number from the total ml./hour to calculate the volume of water/hour. Then multiply both volumes by 24 hours to calculate the total amounts of each.

Sample Problem

Physician order: 1800 kcal. given daily at 3/4 strength

Step 1. 3/4

Step 2. If the kcal./ml. of the house formula is 1.06, divide 1.06 into 1800 to yield 1698 ml. for the total formula volume.

Step 3. Divide 3 into 1698 ml. to yield 566. This is the *factor*.

Step 4. Multiply the factor (566) by 4 to get 2264 ml. This number represents total volume including formula and water.

Step 5. Subtract the 1698 ml. of Step 2 from the 2264 ml. of Step 4 to yield 566 ml. This is the amount of water that must be added to the formula volume as calculated in Step 2 to achieve the strength required.

Note: If the order reads 75 ml./hour, multiply by 3/4 to equal 56.25 ml. of formula/hour. The volume/hour (75 ml.) minus the formula volume/hour (56.25 ml.) = 18.75 ml. water/hour. Totals (× 24 hours) = 1350 ml. of formula mixed with 450 ml. of water = 1800 ml.

METHOD FOR CALCULATION OF TUBE FEEDING STRENGTH USING A POWDERED FORMULA

The same basic principles apply as for adjusting a liquid formula:

Step 1. From the physician's order, determine the fractional strength to prepare, e.g., 1/4, 1/2, or 3/4.
Step 2. Separate the grams of powder and ml. of water in one packet from the total volume/packet.
Step 3. Multiply the grams of powder by the appropriate fraction to yield the adjusted grams of powder/packet.
Step 4. Multiply the original full strength calories/packet times the appropriate fraction to yield the adjusted calories packet.
Step 5. Divide the adjusted calories from Step 4 into the total calories or volume ordered from Step 1 to get a *factor.*
Step 6. Multiply the factor times the adjusted grams in Step 3 to find the total grams needed to achieve the required calories.
Step 7. Divide the total adjusted powder by the grams of powder full strength packet to get the number of packets needed for required calories.
Step 8. Multiply the factor from Step 5 times the ml. of water/packet that yields full strength to produce the total volume of water needed.

Note: If the order reads ml./hour, multiply times 24 hours and begin with Step 1.

Sample Problem

Physician order: 1800 kcal. given daily at 3/4 strength

Step 1. 3/4
Step 2. If the kcal./ml. is one kcal./ml., then the volume is 1800 ml. Otherwise, see previous Sample Problem for liquid formula.

Step 3. The package states that each packet needs 255 ml. of water for full strength. Each standard dilution of one packet is 300 ml. Therefore, each packet must contain 45 grams of powder (300 ml. minus 255 ml.).

Step 4. Multiply 45 grams by 3/4 to yield 34 grams. Each packet contains 300 kcal. in standard dilution times 3/4 to equal 225 kcal./34 grams.

Step 5. Divide 225 kcal. into 1800 kcal. to yield 8. This is the *factor.*

Step 6. Multiply the factor (8) times 34 grams to yield 272 grams.

Step 7. Divide 272 grams by 45 grams to equal 6 packets of powdered formula.

Step 8. Multiply 8 times 255 ml. to yield 2040 ml. of water needed.

Note: Standard dilution is full strength.

REFERENCE NOTES

1. Henderson, Cynthia T. "Safe and Effective Tube Feeding of Bedridden Elderly." *Geriatrics*, 46(8):58, August 1991.

2. Ibid p. 65.

3. Morley, John E. *Geriatric Nutrition.* New York: Raven Press, 1990, pp. 113, 346-347.

X. Documentation of the Medical Record

Most facilities prefer the *Subjective/Objective/Assessment/Plan* (SOAP) method of charting in the Patient Oriented Medical Records (POMR), but the *Data/Assessment/Plan* (DAP) method may be considered. They are very similar, though in the latter the data replaces the *Subjective/Objective* sections. Beginning with progress notes and then care plans, specific points to consider are outlined:

Progress Notes:

1. Should be neat, legible, and specific.
2. Arrange chronologically.
3. Date notes (sometimes the time is also required), identify department, and sign using full name and title after charting (*Example:* 5/1/90 10:00 a.m. Dietetic Service . . . Name, Title)
4. Write in behavioral terms, stating what the patient must be able to do (see under Treatment Planning: Objectives, item 3c).
5. Reflect interventions as stated on the treatment or care plan.
6. State current client condition concerning treatment issues.
7. Document any concerns of or requests made by the client.
8. Document any significant physical/behavioral changes.
9. Document significant outside contact.
10. Document any treatment plan changes made.

Treatment Planning:

1. Problem Statement
   a. Observable and behavioral through patient contact or review of patient's medical records.
   b. Current situation not resolved.

c. Duration or frequency, i.e., how long has it persisted or how often does it occur?
d. Prioritized: state primary problem first, then secondary, tertiary, etc. conditions in order of importance.
e. Mention psychological food-related problem if applicable, e.g., alcoholism, Alzheimer's disease.

2. Goals
   a. State in positive terms.
   b. Will relate to each problem statement or clinical need.
   c. Future-oriented and written in the future tense.
   d. The subject of every goal is the patient.
   e. More general than objectives.
3. Objectives
   a. Measurable in terms of percentages, units, ratios, etc. Refer to 3c.
   b. Stated target date given.
   c. Written in the future tense (i.e., the patient will) and in behavioral terms. Also refer to following paragraph on target behavior.
   d. Steps toward achieving objective are directly related to it.
   e. Achievable, especially during specified time frame.
4. Interventions
   a. The subject of the statement is the staff member.
   b. Who is responsible to do it?
   c. What is to be done?
   d. When is it to be done (frequency)?
   e. Where and how is it to be done (if applicable)?
   f. Related to the problem and to the objective being addressed in order to effectively solve it.
   g. Written in the future tense (i.e., the dietician will).

A target behavior is a behavior that is to be changed or taught. It must be specific (for example: the client *will* consume more orange and grapefruit juices, tomatoes, cantaloupe, and other Vitamin-C rich foods). The staff must be able to agree the behavior has occurred. It must be measurable, usually in mathematical terms (i.e, how much Vitamin-C rich food of which type was eaten). It must be *observable*: the staff must perceive external behavior directly through the five senses (visual, auditory, gustatory, tactile, and olfactory)–a staff member should not presume that he or she understands what a client really thinks or feels. In this case, the Food Intake Study form (see Appendix 10) can be used to record how much Vitamin-C rich food the staff member directly *sees* the client eat at a given meal or snack. The form becomes the tool that communicates this information to the rest of the staff. It must be *objective.* Personal biases or subjective viewpoints are not permissible when evaluating behaviors. In other words, if a staff member does not like grapefruit juice, it must not be assumed the client will not drink it.

When developing progress notes and corresponding care plan, think of yourself as a reporter answering the questions–who, what, when, and sometimes where and how. It is strongly advised that patients with significant weight change, pressure sores, and enteral feedings be documented at least monthly. There is **no substitute** for meal rounds for gaining insight from clients before any documentation occurs. Prepare for care plan meetings in advance by gathering data and identifying problems or needs.

Follow-up progress notes reflect changes in the following items: diagnosis, medical condition, allergies, diet, feeding, chewing and swallowing ability, dental status, skin condition, elimination function, mobility, and mental status. Appetite level, weight status, and new lab data (including accuchecks) or medications (including discontinuations and PRNs) should also be reported. The checklist-style Follow-Up Information form in Appendix 14 alleviates the need for lengthy narrative explanations. However, brief notes may be made on this form, or a Nutrition Progress Notes form (Appendix 15) may be utilized and attached.

Appendixes

Forms in these appendixes are available from Professional Printing & Publishing, Inc. P.O. Box 5758, Bossier City, LA 71171-5758. Toll Free WATS (all states) 1-800-551-8783: FAX 318-746-6995.

Appendix 1. Meal Patterns (Form 500901WA)

	1000	1200	1400	1500
Breakfast–				
Fruit or Juice	1	1	1	1
Starch	1	1	2	2
Meat	1	1	1	1
Margarine	0	1	1	1
Milk	1/2	1/2	1/2	1
Lunch–				
Meat	2	2	2	2
Starch	1	1	1 1/2	2
Vegetable	1	1	1	1
Fruit	1	1	1	2
Margarine	0	0	1	1
Milk	1/2	1/2	1/2	0
Supper–				
Meat	1	2	2	2
Starch	1	1	2	2
Vegetable	1	1	1	1
Fruit	1	1	1	1
Margarine	0	0	0	0
Milk	1/2	1/2	1/2	1/2
H.S. Snack–				
Milk	1/2	1/2	1/2	1/2
Starch	1/2	1	1	1
Meat	0	0	0	0
Fruit or Juice	0	0	0	0
Calories	985	1184	1409	1489
Carbohydrate (gms.)	131.5	139.0	176.5	199.0
Protein (gms.)	58.5	67.0	74.5	76.0
Fat (gms.)	25.0	40.0	45.0	45.0

Calorie and carbohydrate distribution is approximately 2/7, 2/7, 2/7, 1/7 between breakfast, lunch, supper, and h.s. snack. This pattern is planned for 2% milk.

1600	1800	2000	2200	2400
1	2	2	2	2
2	2	2	2	3
1	1	1	1	2
1	1	1	1	1
1	1/2	1/2	1	1
3	3	3	3	3
2	2	2	2	2
1	1	1	1	1
2	2	2	2	2
1	1	1	1	2
0	1/2	1/2	1	1
2	2	3	3	3
2	2	2	2	2
1	1	1	1	1
1	1	1	2	2
0	1	2	2	2
1/2	1	1	1	1
1/2	0	0	0	0
1	1	2	2	2 1/2
0	1	1	1	1
0	1	1	1	2
1578	1816	2006	2191	2404
199.0	229.0	244.0	271.0	308.5
83.0	90.0	100.0	108.0	112.5
50.0	60.0	70.0	75.0	80.0

Appendix 2. Sensory Evaluation Device (Form 500401WA)

Name of Evaluator ________________ Date _____ Day of Week _______							
Meal: ____ Breakfast ____ Lunch ____ Dinner ____ Special Diet __________							
Rate Items Below on a 1 to 10 Scale, 10 Being the Highest, 1 the Lowest							
Items	Appearance	Taste	Aroma	Texture	Color	Temp.	Rating

Appearance	Taste	Aroma	Texture	Color	Temperature
Dry Greasy Liquid Shape Size	Bland/Flat Burnt Chemical Metallic Musty Salty Seasoning Spicy Sweetness	Pungent Unpleasant Weak	Coarse Crumbled Curdled Dry Fine Grainy Moist Soft Thick Thin Tough	Cloudy Dark Light Unnatural	Cold Hot
Rater's Comments:					
Client's Comments:					

Appendix 3. Abbreviations and Equivalents (Form 502501WA)

LF	–	Low fat; indicates omission of butter, lard, and margarine in food preparation and service.
LS	–	Low salt; indicates omission of salt in food preparation and service.
Diet	–	Low calorie salad dressing or low calorie mayonnaise that is less than 16 calories/Tbsp.

(Preference is given to "low" instead of "free" because there are few products that meet the zero level criteria indicated by the term "free.")

Unswt.	–	Unsweetened
Str.	–	Strained or pureed foods
t. or tsp.	–	Teaspoon
T. or Tbsp.	–	Tablespoon
Qt.	–	Quart
Gal.	–	Gallon

dash or speck	=	less than 1/8 tsp.			
1 ounce	=	28.35 cc. or ml.			
1 ounce	=	2 tablespoons			
1 tablespoon	=	3 teaspoons	7/8 cup	=	14 tablespoons
1/8 cup	=	2 tablespoons	1 cup	=	16 tablespoons
1/8 cup	=	1 fluid ounce	1/2 pint	=	1 cup
1/4 cup	=	4 tablespoons	1/2 pint	=	8 fluid ounces
1/3 cup	=	5 tablespoons + 1 tsp.	1 pint	=	2 cups
3/8 cup	=	6 tablespoons	1 quart	=	2 pints
1/2 cup	=	8 tablespoons	1 gallon	=	4 quarts
5/8 cup	=	10 tablespoons	1 peck	=	8 quarts (dry)
2/3 cup	=	10 tablespoons + 2 tsp.	1 bushel	=	4 pecks
3/4 cup	=	12 tablespoons	16 ounces	=	1 pound

Appendix 3 (continued)

Portion Scoop or Disher Number	Level Measure
6	2/3 cup
8	1/2 cup
10	3/8 cup
12	1/3 cup
16	1/4 cup
20	3 1/5 tablespoons
24	2 2/3 tablespoons
30	2 1/5 tablespoons
40	1 3/5 tablespoons
60	1 tablespoon
100	scant 2 teaspoons

Ladle Number	Approximate Measure
1 ounce	1/8 cup
2 ounce	1/4 cup
4 ounce	1/2 cup
6 ounce	3/4 cup
8 ounce	1 cup
12 ounce	1 1/2 cups

Size of Cans	*Average Contents*
#1 1/2 flat	1 cup
Picnic	1 1/4 cup
#300	1 3/4 cup
# 1 tall	2 cups or 1 pint
#303	2 cups or 1 pint
#2	2 1/2 cups
#2 1/2	3 1/2 cups
#3 cylinder	5 3/4 cups
#10	12 cups or 3 quarts

REFERENCE

1. Shugart, Grace and Molt, Mary, *Food for Fifty*, 9th Edition, New York: Macmillan Publishing Co., Inc., 1993.

Appendix 4. Sanitation Checklist (Form 500801WA)

Kitchen Components and Equipment	1	2	3	4	5	6	7	8	9	10	11	12	13
DAILY													
1. Can Opener													
2. Food Processor													
3. Floors (all)													
4. Food Carts													
5. Steam Table													
6. Steamer													
7. Mixers													
8. Ovens													
9. Refrigerators													
10. Sinks													
WEEKLY													
11. Pot and Pan Racks													
12. Cabinets													
13. Trash Cans													
14. Dish Machine: De-liming													
MONTHLY													
15. Dish Machine: Clean Manifolds and Nozzles													
16. Floor and Sink Drains													
17. Pantry Shelves													

Initial appropriate box upon completion of assigned task. Check schedule for assignments.

Month of:

Date of Month																	
14	15	16	17	18	19	20	21	22	23	24	25	26	27	28	29	30	31

Appendix 5. Dietary In-Service Plan (Form 500201WA)

Year ______________ Facility ______________________________________

Month	Title of Program	Instructor
January		
February		
March		
April		
May		
June		
July		
August		
September		
October		
November		
December		

Appendix 6. Training Report (Form 500301WA)

Facility Name: ______________________________

Department: ______________________ Date: ____________

Length of Program: __________________ Time: ____________

Meeting Area: ______________________________

Group(s) required to attend: ______________________

Subjects covered: ______________________________

____

Problems, comments, suggestions: ____________________

____

____

Conducted by: ______________________________

Title(s): ______________________________

Signature(s): ____________________

____

Title(s): ____________________

____

Employee Name	Title	Employee Name	Title

Appendix 7. Consulting Dietitian's Report (Form 500602WA)

I. CLINICAL SECTION

A. Summary of Clinical Documentation (indicate number completed):

____ Initial nutritional assessments ____ Care plans

____ Annual nutritional assessments ____ Care plans

____ Monthly progress notes ____ Care plans

____ Quarterly progress notes ____ Care plans

____ Progress notes due to significant changes as related to diet ____ Care plans

B. Diet instructions ______ Group ______ Individual

II. ADMINISTRATIVE SECTION

A. Meal Observation Checklist: Meal ______ Date ______ Time ______

Substitutions of similar nutrient value offered	______ Yes	______ No
Courteous staff	______ Yes	______ No
Meal served on time	______ Yes	______ No
Self-help device available	______ Yes	______ No
Adequate and timely assistance	______ Yes	______ No
Food temperature acceptable	______ Yes	______ No

Specify temperature and location ________________________________

B. Kitchen Observation Checklist: Date ______ Time ______

Tray accurate	______ Yes	______ No
Tray attractive	______ Yes	______ No
Food covers used	______ Yes	______ No
Serviceware in good repair	______ Yes	______ No
Serviceware clean	______ Yes	______ No
Plate waste minimal	______ Yes	______ No
Proper temperatures in storage areas	______ Yes	______ No

C. Nourishment Observation Checklist: Date ______ Time ______

Bedtime nourishment offered	______ Yes	______ No
Snack or supplement consistent with physician order	______ Yes	______ No
Snack or supplement available per patient request	______ Yes	______ No

III. MANAGERIAL SECTION

A. Menus completed for next food order ______ Yes ______ No

B. Annual in-service scheduled or given
(please attach any training report if yes) ______ Yes ______ No

C. Policies and procedures up to date ______ Yes ______ No

Please specify section and elaborate on any "no" answer.

____

____

____

____

IV. RECOMMENDATIONS

____

____

____

V. COMMENTS

____

____

____

Visitation Dates	Hours

Signature ______________________ Date ______

Administrator's Signature ______________________ Date ______

Kitchen Manager's Signature ______________________ Date ______

Appendix 8. Quality Assurance Program (Form 500701WA)

QUALITY ASSURANCE PROGRAM

Monthy Summary Worksheet

INDICATOR	MET	NOT MET	%	COMMENTS

% is the number met divided by total number done for each indicator.

Signature ______________________________________ Date ___________

Appendix 9. Nutrition Screening (Form 500501WA)

Admission Date ______ Name ______________________ Today's Date ________

Female ____ Male ____ Date of Birth ________ Ht. ______ Wt. ______

Weight Change Past 6 Months ____________ B.P. ________

Diet Order ______________________ Food Intolerance ______________________

Lab values:

Hgb ____ Hct ____ FBS ____ BUN ____ Albumin ____ Cholesterol ____

Major DX ____

____

Medications ____

____

Chewing/Swallowing Problem Yes__No__ Tube Fed Yes__No__

Smoking Hx Yes__No__ Drug/Alcohol Abuse Yes__No__

Parenteral IV Yes__No__ Poor Appetite Yes__No__

PLEASE DO NOT WRITE BELOW THIS SPACE. FOR DIETARY USE ONLY.

IBW ______ If not at IBW, % above or below IBW ______ Pt. age ______

NUTRITION CARE PRIORITY LEVEL

No apparent nutrition-related problem ______

Mild abnormality of one or two nutrition-related parameters ______

Potential for nutrition risk due to diagnosis and/or treatment ______

Nutritionally at risk for multiple reasons ______

Appendix 10. Food Intake Study (Form 501401WA)

NAME: ______________________________

Please enter in the appropriate column how much the patient ate.

DATE:				
BREAKFAST	Fruit/Juice			
	Cereal			
	Entree			
	Toast–Butter			
	Beverage			
LUNCH	Soup/Salad/Juice			
	Entree			
	Starch			
	Vegetable			
	Bread–Butter			
	Dessert			
	Beverage			
SUPPER	Soup/Salad/Juice			
	Entree			
	Starch			
	Vegetable			
	Bread–Butter			
	Dessert			
	Beverage			
	Snacks (Include Time)			

Appendix 11. Calorie Count (Form 501001WA)

Patient Name: ______________________ Ward: __________________
Please enter the date and how much the patient ate in the appropriate boxes.

BREAKFAST				PRO	CHO	FAT	KCAL
Fruit/Juice							
Cereal							
Eggs or Meat							
Toast							
Milk							
DINNER							
Soup/Salad/Juice							
Main Entree							
Starch							
Vegetable							
Bread							
Margarine							
Dessert/Fruit							
Milk							
SUPPER							
Soup/Salad/Juice							
Main Entree							
Starch							
Vegetable							
Bread							
Margarine							
Dessert/Fruit							
Milk							
SNACKS							
Mid-A.M.							
Mid-P.M.							
H.S.							

Appendix 12. Nutritional Assessment (Form 501101WA)

Date ____ Sex ____ Birthdate ____ Height ______ Weight ______ Ibw: ____

Weight History | B.P.

Admitting Diagnosis

Diet Order | Texture | Supplement

Food Allergies: □ Tomatoes □ Eggs □ Citrus □ Chocolate □ Other ________

Food Intolerances: □ Dairy □ Onion/Cabbage Family □ Other ____________

Meds: □ Cardiovascular □ Antiarthritic □ Psychoactive □ Other __________

Swallowing Ability: □ Choking □ Regurgitation □ Aspiration □ Other ______

Functional Feeding Ability: □ Independent □ Assisted □ Total Feed

Adaptive Equipment _______________ Dining Location _______________

Oral Status: □ Edentulous □ Dentures
□ ↑ □ ↓ □ Missing Teeth _________
□ Natural □ Mouth Pain □ Dry Mouth

Sensory Status: Gustatory □ Alteration
Visual □ Glasses □ Cataracts
Auditory □ Hearing Aid □ Impaired

Activity Level:
□ Ambulatory
□ Non-Ambulatory
□ Bedfast
□ Tremors

Cognitive Pattern:
□ Oriented
□ Comatose
□ Incoherent
□ Memory Loss

Bowel and Bladder Function:
□ Regular □ Constipated □ Diarrhea
□ Incontinent □ Catheter
□ Laxative Specify ____________

Skin Condition:
Pressure Sore ____ Yes ____ No
Stage ____ Location ____________
Edema ____ Yes ____ No Degree ___
Location ______________________

Special Risks: □ Fever □ Emesis □ UTI

Intake Level: □ 100% □ 75% □ 50% □ 25%

Fluid Ingested (CC or OZ. Daily) _______

Weight and Lab Data	Initial		Subsequent Results					
Enter date and results	Date	Result	Date	Result	Date	Result	Date	Result
Weight								
Glucose								
Hemoglobin/ Hematocrit								
Potassium								
Albumin								
Cholesterol								

Appendix 12 (continued)

Nutritional Requirements:

Male: BEE = 66 + (_______) + (_______) – (_______) = __________
6.2 × WT.(LB.) 12.7 × HT.(IN.) 6.8 × Age Kilocalories

Female: BEE = 665 + (_______) + (_______) – (_______) = _________
4.4 × WT. (LB.) 4.3 × HT. (IN.) 4.7 × Age Kilocalories

Daily Fluid Requirements: 30 CC./Kg. × _______ Kg. Body Wt. __________ CC.

Total Daily Energy Expenditure (TDE) = ____ × ____ × ______ = _________
BEE Activity Factor Injury Factor Kilocalories

Note: Add/Substract 500 kcal. for 1 lb. gain/loss per week.
Protein Needs = _______ × ______ = _________
IBW (Kg.) Protein Factor Protein (Gm.)

Activity Factor		Injury Factor		Protein Requirement Factor	
Comatose	= 1.1	Minor surgery	= 1.1	Normal	= 0.8-1.0
Bedfast	= 1.2	Major surgery	= 1.2	Pressure Sore, Fracture, Infection Depending Upon Severity	= 1.1-2.0
Sedentary	= 1.3	Skeletal trauma	= 1.35		
Active	= 1.5	Infection	= 1.2 Mild, 1.4 Moderate, 1.6 Severe		

Patient Name ____________ Physician ______________ Room No. _____

Appendix 13. Enteral Assessment (Form 501201WA)

Date ____ Sex ____ Birthdate ____ Height ______ Weight ______ Ibw: ____

Weight History | B.P.

Admitting Diagnosis

Formula Order | Fluid Order

Formula CC. Per Day ______ × ______ Kcal./CC = ______ Kcal. in Formula
Formula CC. Per Day ______ × ______ Gm. Protein/CC. = ____ Pro. in Formula
Formula CC. Per Day ______ × ______ % Formula Fluid = ______ Formula Fluid
Formula Fluid ______ + ______ Fluid Order = ______ Total Fluid per Day (CC.)

Meds: □ Cardiovascular □ Antiarthritic □ Psychoactive □ Other __________

Swallowing Ability: □ Choking □ Regurgitation □ Aspiration □ Other _______

Oral Status: □ Edentulous □ Dentures
□ ↓ □ ↑ □ Missing Teeth ________
□ Natural □ Mouth Pain □ Dry Mouth

Sensory Status: Gustatory □ Alteration
Visual □ Glasses □ Cataracts
Auditory □ Hearing Aid □ Impaired

Skin Condition:
Pressure Sore ____ Yes ____ No
Stage ____ Location ______________
Edema ____ Yes ____ No Degree ___
Location ________________________

Bowel and Bladder Function:
□ Regular □ Constipated □ Diarrhea
□ Incontinent □ Catheter
□ Laxative Specify ______________
__

Weight and Lab Data	Initial		Subsequent Results					
Enter date and results	Date	Result	Date	Result	Date	Result	Date	Result
Weight								
Glucose								
Hemoglobin/ Hematocrit								
Electrolytes (Specify)								
Albumin								
Triglycerides								
TIBC								
BUN								

Appendix 13 (continued)

Nitrogen Balance = (Protein intake ____ in Gm. ÷ 6.25) – (UUN in gm. + 4) = ____ Gm. nitrogen

Nutritional Requirements:

Male: BEE = 66 + (______ 6.2 × Wt.(LB.)) + (______ 12.7 × Ht.(IN.)) – (______ 6.8 × Age) = __________ Kilocalories

Female: BEE = 665 + (______ 4.4 × Wt. (LB.)) + (______ 4.3 × Ht. (IN.)) – (______ 4.7 × Age) = __________ Kilocalories

Daily Fluid Requirements: 30 CC./Kg. × ______ Kg. Body Wt. __________ CC.

Total Daily Energy Expenditure (TDE) = ____ BEE × ____ Activity Factor × ______ Injury Factor = __________ Kilocalories

Note: Add/Substract 500 kcal. for 1 lb. gain/loss per week.

Protein Needs = ______ IBW (Kg.) × ______ Protein Factor = __________ Protein (Gm.)

Activity Factors		Injury Factors		Protein Requirement Factors	
Comatose	= 1.1	Minor surgery	= 1.1	Normal	= 0.8-1.0
Bedfast	= 1.2	Major surgery	= 1.2	Pressure Sore, Fracture, Infection Depending Upon Severity	= 1.1-2.0
Sedentary	= 1.3	Skeletal trauma	= 1.35		
		Infection	= 1.2 Mild, 1.4 Moderate, 1.6 Severe		

Patient Name ____________ Physician ______________ Room No. _____

Appendix 14. Follow-Up Information (Form 501301WA)

Explain Any Yes Answers	Date	Date	Date
Change in Diagnosis (Dx)	Yes __ No __	Yes __ No __	Yes __ No __
Change in Diet/Snack	Yes __ No __	Yes __ No __	Yes __ No __
Change in Preferences or Allergies	Yes __ No __	Yes __ No __	Yes __ No __
Change in Condition (Presence of fever, emesis, U.T.I., water retention, etc.)	Yes __ No __	Yes __ No __	Yes __ No __
Change in Meds	Yes __ No __	Yes __ No __	Yes __ No __
Change in: Swallowing Ability Feeding Ability Adaptive Feeding Device Chewing Ability Vision/Hearing/Speech Mobility/Dining Location Cognitive Pattern	 Yes __ No __ Yes __ No __ Yes __ No __ Yes __ No __ Yes __ No __ Yes __ No __ Yes __ No __	 Yes __ No __ Yes __ No __ Yes __ No __ Yes __ No __ Yes __ No __ Yes __ No __ Yes __ No __	 Yes __ No __ Yes __ No __ Yes __ No __ Yes __ No __ Yes __ No __ Yes __ No __ Yes __ No __
Skin Condition Stable	Yes __ No __	Yes __ No __	Yes __ No __
Appetite Level	Specify:	Specify:	Specify:
Fluid Intake Level	Specify:	Specify:	Specify:
Weight Status	% loss or gain if significant __________	% loss or gain if significant __________	% loss or gain if significant __________
New Lab Values	Yes __ No __	Yes __ No __	Yes __ No __

Patient Name ____________ Physician ______________ Room No. _____

Appendix 15. Nutrition Progress Notes (Form 501501WA)

Date, Signature, and Title must accompany all entries.

Resident Name ____________ Physician ______________ Room No. _____

Index